PRAISE FOR THE FIRST EDITIONS

"A deeply meaningful book and a much-needed corrective to our opinions about the natural world, what membership means—in our bodies and in the Body."
Sheldon Vanauken, author of *A Severe Mercy*

"An enthralling book that I wish I had had the insight to write."
C. Everett Koop, former Surgeon General

"Reflections and stories by a humane tender of wounds. Paul Brand's book, like his life as a doctor, is wholly admirable."
Richard Selzer, author of *Mortal Lessons* and *Confessions of a Knife*

"An unusual medical, biological, and spiritual work. A great appreciation of the human body, life, and belief in God fills this book."
Joseph E. Murray, Nobel Laureate, *The New England Journal of Medicine*

PRAISE FOR THIS EDITION

"In a world divided by races and faces, healers Philip Yancey and Paul Brand knit us together by blazing a magnificent journey through the human body—inspiring discovery, cell by cell, of God's marvel of a blueprint for his unified church. A shining collaboration, their book validates well that if every cell matters, so does every life. Yet only in relation to each other do we come alive. In proving that view, Yancey and Brand achieve an uplifting and awakening balm for today's bruised and fraying church. This book is a glorious achievement, and it arrives kindly with soothing correction. But best of all, it's right on time."
Patricia Raybon, author of *My First White Friend* and *Undivided: A Muslim Daughter, Her Christian Mother, Their Path to Peace*

"I have greatly appreciated my friend Philip Yancey's work over the years. I enthusiastically commend this book in which Philip offers his wisdom and the wisdom of Dr. Paul Brand to a new generation of readers. I anticipate that it will be of great value to Christians and non-Christians alike."
John M. Perkins, cofounder, Christian Community Development Association, president emeritus of the John Perkins Foundation

"I have long loved the collaboration between the marvelous theologian Philip Yancey and his mentor—orthopedic, humanitarian surgeon Paul Brand. What a blessing to reread this work after thirty years, to immerse myself in the story of Dr. Brand's work with leprosy patients in India and the Bayou, and in the resultant lessons about the miraculous workings of the human body, the beauty of human friendship and caring, the profundity of God's love. Brilliant, charming, and wonder inducing, this is a modern classic of science and faith."
Anne Lamott, author of *Traveling Mercies, Bird by Bird*, and *Hallelujah Anyway*

"This book is a testimony to the peculiar mystery of God's presence—the Word made flesh, as John's Gospel names it. For people whose spiritual lives have been too disembodied, Dr. Brand's attention to living bodies invites us to return to our true selves. At the same time, Yancey has rendered enfleshened truth about our existence in words. I pray you can receive this combination of wisdom and craft as the gift that it is—a word in season, inviting each of us deeper into the life that is really life."

Jonathan Wilson-Hartgrove, author of *Reconstructing the Gospel*

"Dr. Brand is a legendary medical doctor, and Philip is a doctor of the soul. Together they have written one of the greatest books on life . . . in a world where far too many lives are disgraced and desecrated. May it make you a champion of life, just as both of these men have been, and may we be death's greatest adversary."

Shane Claiborne, cofounder of The Simple Way and Red Letter Christians, author of *Beating Guns*

"God's imaginative genius with his physical creation always has astounding parallels in the spiritual realm, and it's no different with the epitome of his creation, the human body. In their timeless Christian classic *Fearfully and Wonderfully*, my friend Philip and his coauthor Dr. Brand draw amazing parallels between our bodies and the Body of Christ, which leave the reader startled and breathless with wonder. This remarkable book had a profound impact on my life when I first read it years ago, and its insights will astonish you for sure!"

Joni Eareckson Tada, Joni and Friends International Disability Center

FEARFULLY *and* WONDERFULLY

The MARVEL *of* BEARING GOD'S IMAGE

DR. PAUL BRAND *and*
PHILIP YANCEY

UPDATED AND COMBINED EDITION

An imprint of InterVarsity Press
Downers Grove, Illinois

InterVarsity Press
P.O. Box 1400, Downers Grove, IL 60515-1426
ivpress.com
email@ivpress.com

InterVarsity Press® is the book-publishing division of InterVarsity Christian Fellowship/USA®, a movement of students and faculty active on campus at hundreds of universities, colleges, and schools of nursing in the United States of America, and a member movement of the International Fellowship of Evangelical Students. For information about local and regional activities, visit intervarsity.org.

While any stories in this book are true, some names and identifying information may have been changed to protect the privacy of individuals.

Published in association with Creative Trust Literary Group LLC, 210 Jamestown Park Drive, Suite 200, Brentwood, TN 37027, www.creativetrust.com.

Cover design and image composite: David Fassett
Interior design: Daniel van Loon
Images: gold texture background: © Katsumi Murouchi / Getty Images
    blue watercolor: © zoom-zoom / iStock / Getty Images Plus
    abstract oil painting: © boonchai wedmakawand / Getty Images
    marble pattern: © Alexey Bykov / iStock / Getty Images Plus
    fingerprints: © CSA Images / Getty Images
    human anatomy: © CSA Images / Vetta / Getty Images
    abstract circle brush stroke: © wacomka / iStock / Getty Images Plus

ISBN 978-0-8308-4570-5 (print)
ISBN 978-0-8308-6568-0 (digital)

Printed in the United States of America ∞

Library of Congress Cataloging-in-Publication Data
Names: Brand, Paul W, author. | Yancey, Philip, author.
Title: Fearfully and wonderfully : the marvel of bearing God's image / Dr.
   Paul Brand and Philip Yancey.
Description: Updated and Combined Edition. | Downers Grove : InterVarsity
   Press, 2019. | Combines and updates two previous titles: Fearfully and
   wonderfully made and In his image.
Identifiers: LCCN 2019019326 (print) | LCCN 2019021529 (ebook) | ISBN
   9780830865680 (eBook) | ISBN 9780830845705 (hardcover : alk. paper)
Subjects: LCSH: Church. | Human body--Religious aspects--Christianity.
Classification: LCC BV600.3 (ebook) | LCC BV600.3 .B725 2019 (print) | DDC
   262--dc23
LC record available at https://lccn.loc.gov/2019019326

P	23	22	21	20	19	18	17	16	15	14	13	12	11	10	9	8	7	6	5	4	3	2
Y	38	37	36	35	34	33	32	31	30	29	28	27	26	25	24	23	22	21	20	19		

You created my inmost being;

you knit me together in my mother's womb.

I praise you because I am fearfully and wonderfully made.

PSALM 139:13-14

CONTENTS

A NEW EDITION for a NEW TIME

Philip Yancey

I FIRST LEARNED ABOUT DR. PAUL BRAND in 1976 while writing my book *Where Is God When It Hurts.* As I was pondering the problem of pain, my wife found in the closet of a medical-supply house an intriguing essay he had written on "The Gift of Pain." Dr. Brand had a unique point of view: while most people seek to escape pain, he had spent several million dollars in an effort to *create* a pain system. "Thank God for pain!" he wrote. "I cannot think of a more valuable gift for my leprosy patients."

After training as an orthopedic surgeon in England, Dr. Brand spent most of his medical career in India, where he made a dramatic discovery about leprosy, one of the oldest and most feared diseases. Careful research convinced him that the terrible manifestations of that cruel disease—missing toes and fingers, blindness, skin ulcers, facial deformities—all trace back to the single cause of painlessness. Leprosy silences nerve cells, and as a result its victims unwittingly destroy themselves, bit by bit, because they cannot feel pain.

When Dr. Brand moved from India to a high-tech laboratory in Louisiana, he applied what he had learned about insensitivity and painlessness to diseases such as diabetes. Former Surgeon General C. Everett Koop later told me that Dr. Brand's findings revolutionized the treatment of diabetic feet, helping prevent tens of thousands of amputations each year.

Dr. Brand's work earned accolades on several continents. Queen Elizabeth II appointed him Commander of the Order of the British Empire, India's Mahatma Gandhi Foundation selected him as the only Westerner to serve on that board, and the US Public Health Service gave him their highest award. Despite such international recognition, humility struck me as his strongest attribute.

When I met him, Dr. Brand was still adjusting to life in the United States. Everyday luxuries made him nervous, and he longed for a simple life close to the soil. He preferred going barefoot and spent his spare time bird-watching and tending his garden. Although he knew people such as Gandhi, Mother Teresa, Albert Schweitzer, and Prince Philip, he rarely mentioned them. He talked openly about his failures and always tried to deflect credit for his successes to his colleagues. Most impressively to me, the wisest and most brilliant man I have ever known devoted much of his life to some of the lowliest people on the planet: members of India's Untouchable caste (now called Dalits) afflicted with leprosy.

CONTINUING THE LEGACY

The conversations that stand out sharpest to me now are those in which Dr. Brand recalled individual patients, "nobodies" on whom he had lavished medical care. When he began his pioneering work, he was the only orthopedic surgeon in the world working among fifteen million victims of leprosy. He and his wife, Margaret, performed several dozen surgical procedures on some of these patients, transforming rigid claws into usable hands through innovative tendon transfers, remaking feet, forestalling blindness, transplanting eyebrows, fashioning new noses.

He told me of the patients' family histories, the awful rejection they had experienced as the disease presented itself, the trial-and-error treatments of doctor and patient experimenting together. Almost always his eyes would moisten and he would wipe away tears as he remembered their suffering. To him these people, among the most neglected on earth, were not nobodies but persons made in the image of God, and he dedicated himself to honor and help restore that image.

As I got to know him, Dr. Brand admitted to me somewhat shyly that he had once attempted a book. After hearing him deliver a series of talks to the Christian Medical College in India, other faculty members urged him to write them down for publication. The result filled only ninety pages, not enough for a book. Twenty years had passed, and he had not touched the manuscript since. I persuaded him to dig through closets and bureau drawers until he located the badly smudged third carbon copy of those chapel talks, and that night I sat up long past midnight reading his remarkable meditations on the human body.

Dr. Brand described his goal in writing:

> In a sense we doctors are like employees at the complaint desk of a large department store. We tend to get a biased view of the quality of the product when we hear about its aches and pains all day. In this little manuscript, I tried instead to pause and wonder at what God made: the human body.

In a further step, Dr. Brand lifted an analogy from the New Testament, the Body of Christ, and updated it with his knowledge from modern science.

I asked for, and graciously received, the freedom to take his original manuscript and develop it, adding many stories from his life while expanding the medical and spiritual insights. For nearly a decade I followed Dr. Brand around the world, retracing the steps of his medical training in England and observing him with patients at leprosariums in India and Louisiana. Sitting under a tamarind tree, watching him at work in a laboratory, bouncing along in a Land Rover, screeching through the London "tube" (subway), I asked him every question I could think of.

Even after forty years of medical practice, he retained a boyish enthusiasm for the grandeur of the human body. Our conversations roamed wide, yet every topic I brought up, he had already thought about in some depth. He quoted Shakespeare and discussed the derivation of Greek, Hebrew, and Latin words. During breaks he taught me such things as how to select a ripe fig (watch the butterflies) and how weaver birds build their elaborate nests using only one foot and a beak. As a young writer, I greedily absorbed his wisdom.

In the years since, some 700,000 readers have bought copies of *Fearfully and Wonderfully Made* or its sequel, *In His Image*. I have heard from many: pregnant women thanking us for helping them appreciate the miracle of life and birth, medical students crediting these books for drawing them toward a career in medicine, high school and college biology teachers who use excerpts in their classes, and other readers grateful for a reminder of the marvels of the human body.

Eminent physicians have also written in praise of Dr. Brand's legacy: Joseph Murray, the Nobel Laureate who performed the first kidney transplant; the medical authors Richard Selzer and Abraham Verghese; the neurosurgeon Oliver Sacks; Dame Cicely Saunders, founder of the modern hospice movement. Although not all of them shared his Christian faith, they found wisdom in his words and inspiration in his humane style of medicine.

UPDATED AND REVISED

Both books remained in print for more than three decades, and their age was beginning to show. After hearing from so many enthusiastic readers over the years, I wanted to introduce Dr. Paul Brand to a new generation. Since medicine and science have advanced greatly, I have made editorial revisions and updated relevant details, in the process condensing the text to create this new, combined edition.

Although Dr. Brand died in 2003, I have retained his first-person voice from the days when he lived in India and in Louisiana. He reflects on his life from the rich vantage point of scientist, missionary, surgeon, storyteller, and theologian.

We live in divided times. Politically, racially, and religiously, the United States is experiencing a severe strain on its unity, and a similar factiousness has spread around the world. We have much to learn from a man who studied medicine during the blitz of World War II, when virtually the entire world was at arms, and who began his medical career during the Partition of India, a cataclysmic event that caused more than a million deaths and created fourteen million refugees.

The modern world depends on institutions—global alliances, corporations, church denominations, government bureaucracies—many of which seem to be failing. Perhaps it is time to take a look at a different kind of community, not an *organization* but rather an *organism*, best illustrated in the human body.

Modern culture has also become reductionistic. We reduce music, movies, and knowledge to blips of data that can be stored on a handheld smartphone. Science, medicine, and other endeavors exist in professional silos with little cross-fertilization. Dr. Brand had the rare ability to bring fields together. An avid scientist, bird-watcher, mountain climber, and organic gardener, he strove to integrate the natural order with the spiritual order. As a Christian, he delighted in discovering echoes of the Creator within the world of nature.

If you train a telescope on the galaxies, stars, and planets of the universe, and then look through a powerful microscope at tiny molecules and atoms and electrons, you will notice an unmistakable similarity in structure and pattern. The same Creator designed both levels of reality. So too the same Creator designed the human body and then inspired New Testament authors to look there for a model of spiritual truth. Dr. Brand's insights come not as sermons but as observations of how cells work together in community and what we can learn from them.

I hope that this book will help span the chasm that for too long has separated the created world from its Source. God invented matter, investing great creativity in this world and especially in the design of our bodies. The least we can do is be grateful.

I look back with nostalgia on the decade of our collaboration when my own writing career was just beginning. I had splendid subjects to work with: the realm of nature and specifically the magnificent human body; the exotic life of a surgeon who brought healing to people who knew no physical pain and therefore much suffering; and the mystery of Christ's Body, surely the most perilous venture ever made by God, entrusting the divine reputation to the likes of us, God's fickle followers.

A PERSONAL NOTE

True friends get their measure, over time, in their effect on you. As I compare the person I was on our first meeting and the person I am now, I realize that large changes have occurred within me, with Paul Brand responsible for many of them.

Elsewhere, I have written honestly about my early struggles with faith, due in large part to exposure to toxic churches. I can imagine God gently steering me to Dr. Brand (through my wife's serendipitous discovery in a closet, of course) at a critical time in my spiritual journey. *OK, Philip, you've seen some of the worst the church has to offer. Now I'll show you one of the best.*

Paul Brand was both a good and a great man, and I am forever grateful for the time we spent together. My faith grew as I observed with a journalist's critical eye a person enhanced in every way by his faith. No one has affected me more, and I know no one who better illustrates Jesus' most-quoted statement in the Gospels, that "whoever loses his life for my sake will find it." From the perspective of a success-obsessed culture, an orthopedic surgeon devoting his career to some of the poorest and most oppressed people on the planet is an example of "losing his life." Yet Dr. Brand lived as full and rich a life as anyone I know, one that combined professional achievement with enduring qualities of humility and a grand sense of adventure.

As much as anyone, he helped set my course in outlook, spirit, and ideals. I look at the natural world and environmental issues largely through his eyes. From him I also gained assurance that the Christian life I had heard in theory can actually work out in practice. It is indeed possible to live in modern society, achieve success without forfeiting humility, serve others sacrificially, and yet emerge with joy and contentment. Whenever I doubt that, I think back on my time with Paul Brand.

There was an exchange at work in our writing collaboration, I now see. Wounded by the church, plagued by doubts, I had neither the maturity nor the ability to express much of my own fledgling faith. Yet I could write with utter integrity about Dr. Brand's faith, and through that process his words and thoughts became mine too. I now view the ten years I spent working with him as an important chrysalis stage. As a journalist, I gave words to his faith. In exchange, he gave faith to my words.

Simone Weil once said, "Imaginary evil [such as that portrayed in books and movies] is romantic and varied; real evil is gloomy, monotonous, barren, boring. Imaginary good is boring; real good is always new, marvelous, intoxicating." I saw real goodness in Paul Brand and found it indeed marvelous and intoxicating. I feel privileged, as his coauthor, to have had some role in shining a light on his life.

PART ONE

IMAGE BEARERS

What a piece of work is a man! how noble in reason! how infinite in faculties! in form and moving how express and admirable! in action how like an angel, in apprehension how like a god!

WILLIAM SHAKESPEARE

CHAPTER ONE

INVISIBLE MADE VISIBLE

A CURTAIN SCREENED MY GROUP of interns and medical students from the rest of the forty-bed ward at a training hospital in Vellore, India. Activity throbbed in the ward: nurses tending to other patients, families bringing in home-cooked food. Inside the curtain, though, we were giving full attention to our young colleague as he made a diagnosis.

He was half-kneeling, in the posture I had taught him, with his warm hand slipped under the sheet and resting on the female patient's bare abdomen. While his fingers probed gently for telltale signs of distress, he pursued a line of questioning that showed he was weighing the possibility of appendicitis against an ovarian infection. Suddenly something caught my eye—a slight twitch of movement on the intern's face. Was it the eyebrow arching upward? A vague memory stirred in my mind, but one I could not fully recall.

The intern's questions were leading into a delicate area, especially for demure Hindu society. *Had she ever been exposed to a venereal infection?* He looked straight into the woman's eyes as he questioned her in a soothing tone of voice. Somehow his facial expression combined sympathy, inquisitiveness, and warmth. His very countenance coaxed the woman to put aside the awkwardness and tell us the truth.

At that moment my memory snapped into place. Of course! The left eyebrow cocked up and the right one trailing down, the head tilted to one side, the twinkling eyes, the wry, enticing smile—these were unmistakably the features of Professor Robin Pilcher, my old chief surgeon in London. I sharply sucked in my breath.

Startled by my reaction, the students all looked up. I could not help it, for it seemed as if the intern had studied Professor Pilcher's mannerisms and was now replicating them in an acting audition. I had to explain myself. "That is the face of my old chief! What a coincidence—you have *exactly* the same facial expression even though you've never been to England and Pilcher has never visited India."

At first the students stared at me in confused silence. Finally, two or three of them grinned. "We don't know any Professor Pilcher," one said. "But, Dr. Brand, that was *your* expression he was wearing."

Later that evening, alone in my office, I reviewed my time under Pilcher. I had learned from him many techniques of surgery and diagnostic procedures. Evidently, he had also imprinted his instincts, his expression, his very smile, in a way that could be passed down to others. It was a kindly smile, perfect for disarming a patient's embarrassment in order to tease out important clues.

Now I, Pilcher's student, had become a link in the human chain, a carrier of his wisdom to students some nine thousand miles away. The Indian intern, young and brown-skinned, speaking in Tamil, had no obvious resemblance to Pilcher or to me. Yet he had conveyed the likeness of my old chief so precisely that it had transported me back to university days with a start. The experience in that ward gave me a crystalline insight into the concept of "image."

A MYSTERIOUS PHRASE

In modern times the word *image* may connote nearly the opposite of its original meaning. Today, a politician hires an image-maker, a job applicant dresses to present an image of confidence and success, a corporation seeks just the right image in the marketplace. I wish to return to the word's original meaning: a true likeness, not a deceptive illusion.

Think of a ten-pound bundle of protoplasm squirming fitfully in a blanket. The baby's father weighs twenty times as much, with his body parts in different proportions. Yet the mother announces proudly that the baby is the "spitting image" of his father. A visitor peers closely. Yes, a resemblance does exist, evident now in a dimple, slightly flared nostrils, a peculiar earlobe. Before long, mannerisms of speech and posture and a thousand other mimetic traits will bring the father unmistakably to mind.

Such usages of *image*—a baby, a professor's facial expression—shed light on a mysterious phrase from the Bible: the image of God. That phrase appears in the very first chapter of Genesis, and its author seems to stutter with excitement, twice repeating an expression from the preceding verse:

> So God created human beings in his own image.
>> In the image of God he created them;
>> male and female he created them. (Genesis 1:27 NLT)

The very first humans received the image of God, and in some refracted way each one of us possesses this sacred quality.

After each stage of creation, God pronounced it "good." Still, something was lacking until God decided, "Let us make human beings in our image, to be like us" (Genesis 1:26 NLT). Among all earthly creatures, only humanity receives the image of God. But how can visible human beings express a likeness to God, who is invisible spirit?

We share with the animals a body composed of bone, organs, muscle, fat, and skin—and in truth, we fall short compared to the abilities of some animals. A horse easily outruns us; a hawk sees far better; a dog detects odors and sounds imperceptible to us. The sum total of our sheer physical qualities is no more godlike than a cat's. And yet, *we* are made in God's image, with our bodies serving as its repository.

Like a growing child absorbing traits from his parents, like a student learning from his professor, we can take on God's qualities—compassion, mercy, love, gentleness—and express them to a needy, broken world. As spirit, God remains invisible, relying on us to make that spirit visible.

It is a supreme mystery that God has chosen to convey likeness through millions of ordinary people like us. We bear that image collectively, as a Body, because any one of us taken individually would present an incomplete image, one partly false and always distorted, like a single glass chip hacked from a mirror. Yet in all our diversity we can come together as a community to bear something of God's image in the world.

LEARNING FROM THE BODY

I close my eyes, blocking the view outside. Barefoot, I am wiggling the small bones in my right foot, half the width of a pencil and yet strong enough to support my weight in walking. I cup my hand over my ear and hear the familiar "seashell" phenomenon, actually the sound of blood cells rushing through the capillaries in my head. I stretch out my left arm and try to imagine the chorus of muscle cells expanding and contracting in concert. I rub my finger across my arm and feel the stimulation of touch cells, 450 of them packed into every one-inch-square patch of skin.

Inside, a multitude of specialized cells in my stomach, spleen, liver, pancreas, and kidneys are working so efficiently that I have no way of perceiving their presence. All the while, fine hairs in my inner ear monitor a swishing fluid, ready to alert me if I suddenly tilt off-balance.

When my cells work well together, I'm hardly conscious of their existence. Instead, I feel the composite of their activity known as Paul Brand. My body, composed of many parts, is one. And that is the root of the analogy we shall explore.

I picture the human body as a community made up of individual cells. The white blood cell, for example, closely resembles an amoeba, though it possesses far less autonomy. A larger organism determines its duties, and it must sometimes sacrifice its life for the sake of that organism. Yet the white cell performs a singularly vital function. The amoeba flees danger; the white cell moves toward it. A white cell can keep alive a person like Newton or Einstein—or you and me.

The cell is the basic unit of an organism; it can either live for itself, or it can help form and sustain the larger being. The same principle applies to human groups, such as neighborhood communities and even nations. "Ask not what your country can do for you," President John F. Kennedy challenged Americans, "ask what you can do for your country." Membership has its privileges and also its conditions.

The apostle Paul explored this analogy in 1 Corinthians 12, a passage that compares the church to the human body. His analogy takes on even more meaning to me because I deal with the body's cells every day. Following Paul's analogy, I augment it like this:

> The body is one unit, though it is made up of many cells, and though all its cells are many, they form one body. . . . If the white cell should say, "Because I am not a brain cell, I do not belong to the body," it would not for that reason stop being part of the body. And if the muscle cell should say to the optic nerve cell, "Because I am not an optic nerve, I do not belong to the body," it would not for that reason cease to be part of the body. If the whole body were an optic nerve cell, where would be the ability to walk? If the whole body were an auditory nerve, where would be the sense of sight? But in fact God has placed the cells in the body, every one of them, for a reason. If all cells were the same, where would the body be? As it is, there are many cells, but one body.

That analogy conveys a more precise meaning to me because though a hand or foot or ear cannot have a life separate from the body, a cell does have that potential. It can live in the body as a loyal member, or it can cling to its own autonomy. Some cells do enjoy the body's benefits selfishly while maintaining complete independence—we call them parasites or cancer cells. From the human body, we learn important lessons on how to bear God's image.

CHAPTER TWO

HUMAN MIRRORS

I STUDIED MEDICINE IN LONDON during some of the darkest days of World War II. Waves of *Luftwaffe* bombers filled the sky, their engines growling like unbroken thunder, their bomb bays releasing cargoes of destruction. During one period, as many as fifteen hundred planes attacked our city on fifty-seven consecutive nights, with some raids lasting eight hours without pause. Only one thing gave us hope: the courage of Royal Air Force pilots who rose in the skies to battle the Germans.

If anything, Winston Churchill understated our gratitude when he said, "Never in the history of human conflict has so much been owed by so many to so few." As a fire-watcher, I followed the aerial confrontations from my post on a rooftop. RAF Hurricanes and Spitfires, tiny and maneuverable, looked like mosquitoes pestering the huge German bombers. Although their cause seemed futile, and more than half of their planes got shot down, the RAF pilots never gave up. Each night they sent a few more of the dreaded bombers cartwheeling in flames toward the earth, to the cheers of us spectators.

Eventually, Hitler decided that Germany could not sustain further losses from the increasingly accurate fighter pilots, and London slept again. I cannot exaggerate the adoration that Londoners gave to those brave RAF pilots. They were the cream of England, the brightest, healthiest, most confident, and often the handsomest men in the entire country. When they walked the streets in their decorated uniforms, the population treated them as gods. Young boys ran up to touch and ogle their heroes. Women envied the few fortunate enough to walk beside a man in air force blue.

I came to know some of these young men in far different circumstances. The Hurricane fighter plane, though agile and effective, had one fatal design flaw. Its single propeller engine, mounted in front, was fed by a fuel line that ran alongside the cockpit, and a direct hit from a German fighter would cause the cockpit to erupt in an inferno of flames. The pilot could eject, but in the few seconds it took to find the lever, heat would melt off every feature of his face: his nose, his eyelids, his lips, often his cheeks. I met these RAF heroes wrapped in bandages as they began the torturous series of surgeries required to refashion their faces.

I helped treat the damaged hands and feet of the downed airmen, even as a team of plastic surgeons went to work on their burned faces. An airman commonly endured twenty to forty surgical procedures. Sir Archibald McIndoe and his plastic surgeons performed miracles of reconstruction, inventing many new procedures along the way. Throughout that tedious process, morale remained surprisingly high among the pilots, who were fully aware of their patriotic contribution. The pilots downplayed the pain and teased each other about their elephant-man features. They were ideal patients.

FACES FOREVER CHANGED

Gradually, however, as the last few weeks of recuperation neared, a change set in. Many of the pilots asked for minor alterations: a nose flap tucked in just a bit, a mouth turned up some at the corner, a slight thinning of the skin transplant on the right eyelid. The realization dawned on all of us, including the patients, that they were stalling. They simply could not face the world outside the hospital.

Despite the miracles wrought by McIndoe's team, each airman's appearance had changed irreparably. No surgeon could restore the protean range of expression of a handsome young face. You cannot appreciate the flexible, nearly diaphanous tissue of the eyelid until you try to fashion one out of coarser skin from the abdomen. That bulgy, stiff tissue will protect the eye adequately, but without beauty. Although technically a good piece of work, the airman's new face was essentially a scar.

I especially remember an RAF pilot named Peter Foster, who confessed his mounting anxiety as release day approached. "The mirror is the key," he told me. For months you use the mirror as an objective measuring device to gauge the progress your surgeons have made. You study scar tissue, the odd wrinkling of the skin, the thickness of lips, and shape of the nose. You ask for certain adjustments, and the doctors decide whether your request is reasonable.

Near the day of release, your view of the mirror changes. Now, as you gaze at the reflection of a new face—not the one you were born with but an inferior imitation—you begin to see yourself as strangers will see you. In the hospital you have been a hero. On the outside, you will be a freak. Fear creeps in. Will any woman dare to marry that face? Will anyone give it a job?

"As you prepare to enter the world with your new face," said Foster, "only one thing matters: how your family and intimate friends will respond. The surgeons' skill in remaking the face counts for little. What matters is how family members react to the news that the surgeons have done all they can. At that critical moment, do you sense a flicker of awkward hesitation or the assurance that they love and accept you regardless?"

Psychologists followed the airmen's progress. Some had girlfriends or wives who could not accept the new faces and quietly stole away or filed for divorce. These airmen tended to stay indoors, refusing to venture outside except at night; they looked for work they could do at home, alone. In contrast, those whose wives and girlfriends and families stuck by them went on to successful careers and leadership positions—they were, after all, the elite of England.

Peter Foster belonged to the fortunate group. His girlfriend assured him that nothing had changed except a few millimeters' thickness of skin. She loved *him*, not just his face, she said. The two got married shortly before Peter left the hospital.

Not everyone reacted so positively. Many adults looked away quickly when they saw him. Children, cruel in their honesty, made faces and mocked him. Peter wanted to cry out, "Inside I am the same person you knew before! Don't you recognize me?" Instead, he learned to turn toward his wife. "*She* became my mirror. She gave me a new image of myself," he said. "Even now, regardless of how I feel, when I look at her, she gives me a warm, loving smile. That tells me I am OK."

QUASIMODO COMPLEX

Two decades after I worked with the airmen, I read a fascinating article, "The Quasimodo Complex," in the *British Journal of Plastic Surgery*. Two physicians reported in 1967 on a landmark study of eleven thousand prison inmates who had committed murder, prostitution, rape, and other serious crimes. Medicine has long known that emotional conflict may produce physical illness. These doctors raised the possibility of the reverse syndrome: physical deformity may lead to emotional distress that results in crime.

According to the article, 20 percent of adults have surgically correctable facial deformities (protruding ears, misshapen noses, receding chins, acne scars, birthmarks, eye deformities). The researchers found that fully *60 percent* of the eleven thousand offenders had such deformities.

The authors, who named the phenomenon after Victor Hugo's *Hunchback of Notre Dame*, raised disturbing questions. Had these criminals encountered rejection and bullying from school classmates because of their deformities? And could the cruelty of other children have bred in them a response of revenge hostility that later led to criminal acts?

The authors proposed a program of corrective plastic surgery for prison inmates. If society rejects some members because of physical appearance, they reasoned, perhaps altering that appearance will change how they are treated and thus how they behave. Whether a murderer on death row or a pilot in the RAF, a person forms a self-image based largely on what kind of image other people mirror back.

The report on Quasimodo prisoners reduces to statistics a truth that every burn victim and disabled person knows all too well. We humans give inordinate regard to the physical body, or shell, that we inhabit. It takes a rare person indeed, someone like Peter Foster's wife, to look past that outer shell and see the true person inside.

As I ponder the Quasimodo complex, I realize that I too have judged and labeled people by their external appearances. I think back to a family tradition I used to practice with my children. Each year during summer holidays I invented a serial adventure story that featured every member of the family by name and character. I tried to weave in some useful aspects: the children heard accounts of themselves showing courage and unselfishness that perhaps they would aspire to in real life.

The plots also included villains who lured the children into scary situations from which they would have to extricate themselves. As I now recall these stories, I remember with a wince that my chief villains, who reappeared in the stories annually, bore the names Scarface and Hunchback. The two antiheroes would attempt to disguise their features, but sooner or later one of the children would see through the disguises and unmask the villains.

Why, I now wonder, did I give the bad characters in my stories those names and characteristics? I was following the stereotype of equating ugly with bad and beautiful with good. In so doing, did I encourage my own children to equate ugliness with evil, potentially making it harder for them to love people with deformities or scars?

After reading about the Quasimodo complex and recognizing the bias in my own storytelling, I started noticing how culture influences our standards for human worth and acceptance. If we judged the American population by the images that appear in magazines and on television, we would conclude we live in a society of perfect bodies. I have seen how the glorification of physical perfection affects my leprosy patients, who will never achieve the status of an Olympic athlete or a beauty queen.

And I have watched more subtle forces at work in my own children during their school days. A child who is clumsy, shy, or unattractive makes an easy target for mocking and bullying. The "mirrors" around us help determine our images of ourselves. How many Salks or Pasteurs have been lost because of rejection by classmates or colleagues?

I have spent my medical career trying to improve the "shells" of my patients. I strive to restore damaged hands, feet, and faces to the body's original design specifications. I feel deep satisfaction when I see those patients relearn how to walk and use their fingers and then return to their communities with an opportunity for a normal life. Nevertheless, I have an increasing awareness that the physical shell I work on does not constitute the whole person.

My patients are not mere collections of tendons, muscles, hair follicles, nerve cells, and skin cells. Each of them, regardless of deformed appearance and physical damage, is a vessel of God's image. Their physical cells will one day rejoin the basic elements of earth, the *humus*. I believe their souls will live on, and my effect on those souls may have far more significance than all my attempts to improve their physical bodies.

Although I live in a society that honors strength, wealth, and beauty, God has placed me among leprosy patients who are often weak, poor, and unattractive. In that environment, I have learned that all of us are, like Peter Foster's wife, mirrors. Each of us has the potential to summon up in others the spark of godlikeness in the human spirit. Or we can disregard that image and judge on the basis of external appearance.

Mother Teresa used to say that when she looked into the face of a dying beggar in Calcutta, she prayed to see the face of Jesus so that she might serve the beggar as she would serve Christ. In an often quoted passage, C. S. Lewis expressed a related thought:

> It is a serious thing to live in a society of possible gods and goddesses, to remember that the dullest and most uninteresting person you talk to may one day be a creature which, if you saw it now, you would be strongly tempted to worship, or else a horror and a corruption such as you now meet, if at all, only in a nightmare. All day long we are, in some degree, helping each other to one of these destinations.

PERSON TO PERSON

My mailbox bulges with appeals from organizations involved in feeding the hungry, clothing the poor, visiting prisoners, housing refugees, combating sexual trafficking, healing the sick. They describe the desperate conditions of a broken world and request my money to help relieve the pain. Often I give because I've worked among such needs and I know these organizations do indeed spread love and compassion. It saddens me, though, that the only thread connecting millions of donors to that world is the impersonal medium of direct mail. Ink stamped on paper, email blasts, stories and photos edited to achieve the best results—these offer no personal connection.

If I only express love vicariously through a check or credit card, I will miss the person-to-person response that hands-on love summons up. Not all of us can serve in parts of the world where human needs abound. But all of us can visit prisons and homeless shelters, bring meals to shut-ins, and minister to single parents or foster children. If we choose to love only in a long-distance way, *we* will be deprived, for love requires direct contact.

The best illustration of this truth is Jesus Christ, the "express image" of God living on this planet. The book of Hebrews sums up his experience on earth by declaring that we now have a leader who can be *touched* with the feelings of our weaknesses (Hebrews 4:15). God saw the need to come alongside us, not simply love us at a distance.

Jesus only affected a small area of the world, however. In his lifetime he had no impact on the Celts or the Chinese or the Aztecs. Rather, he set in motion a mission that was to spread throughout the world, responding to human needs everywhere. Although we cannot change everything in the world, together we can strive to fill the earth with God's presence and love. When we stretch out a hand to help, we stretch out the hand of Christ's Body.

I was privileged to know Mother Teresa, who was awarded a Nobel Peace Prize for her work in Calcutta among members of India's lowest castes. Her order of sisters sought out the sick and dying in the streets and garbage dumps of Calcutta's alleys, and among these were beggars deformed by leprosy. Several times I consulted with her on the proper treatment of the disease.

When her followers in the Missionaries of Charity find beggars in the street, they bring them to the hospital and surround them with love. Smiling women dab at their sores, clean off layers of grime, and swaddle them in soft sheets. The beggars, often too weak to talk, stare wide-eyed at this seemingly misdirected care. Have they died and gone to heaven?

Why this sudden outpouring of love and the warm, nutritious broth being gently spooned into their mouths?

A reporter in New York once confronted Mother Teresa with those very questions. He seemed pleased with his journalistic acumen. Why indeed should she expend her limited resources on people for whom there was no hope? Why not attend to people worthy of rehabilitation? What kind of success rate could her clinic show when most of its patients died in a matter of days or weeks?

Mother Teresa stared at him in silence, absorbing the questions, trying to comprehend what kind of a person would ask them. She had no answers that would make sense to him, so she said softly, "These people have been treated all their lives like dogs. Their greatest disease is a sense that they are unwanted. Don't they have the right to die like angels?"

Another journalist, Malcolm Muggeridge, struggled with the same questions. He observed firsthand the poverty of Calcutta and returned to England to write about it with fire and indignation. But, he comments, the difference between his approach and Mother Teresa's was that he returned to England while she stayed in Calcutta. Statistically, he admits, she did not accomplish much by rescuing stragglers from a sump of human need. He concludes with the statement, "But then Christianity is not a statistical view of life."

Indeed it is not. Not when a shepherd barely shuts the gate on his ninety-nine before rushing out, heartbroken and short of breath, to find the one that's missing. Not when a laborer hired for only one hour receives the same wage as an all-day worker (Matthew 20:1-16). Not when one rascal decides to repent and ninety-nine upstanding citizens are ignored as all heaven erupts in a great party (Luke 15:4-7). God's love, *agape* love, is not statistical either.

PART TWO

ONE *and* MANY

*We often think that when we have completed
our study of one, we know all about two,
because "two" is "one and one." We forget
that we have still to make a study of "and."*

SIR ARTHUR EDDINGTON

$\mathscr{A}$ ROLE *in the* BODY

As a boy growing up in India, I idolized my missionary father, who responded to every human need he encountered. Only once did I see him hesitate to help: when I was seven, and three strange men trudged up the dirt path to our mountain home.

At first glance these three seemed like hundreds of other strangers who came to our home for medical treatment. Each man wore a breech-cloth and turban, with a blanket draped over one shoulder. As they approached, however, I noticed telltale differences: a mottled quality to their skin, thickened foreheads and ears, feet bandaged with strips of blood-stained cloth. As they came closer, I noticed they also lacked fingers and one had no toes, his legs ending in rounded stumps.

Something ominous was happening, and I didn't want to miss it. After calling my father I scrambled on hands and knees to a nearby vantage point. My heart pounded as I saw a look of uncertainty, almost fear, pass across my father's face. I had never seen that expression on my father.

The three men prostrated themselves on the ground, a common Indian custom that my father disliked. "I am not God—he is the One you should worship," he would usually say and lift the Indians to their feet. Not this time. He stood still. Finally, in a sad voice he said, "There's not much we can do. I'm sorry. Wait where you are, don't move. I'll do what I can."

He ran to the dispensary while the men squatted on the ground. Soon he returned with a roll of bandages, a can of salve, and a pair of surgical gloves he struggled to put on. This confused me, as he seldom wore gloves while treating a patient. Father washed the strangers' feet, applied ointment to their sores, and bandaged them. Strangely, they did not wince or cry out as he cleaned and wrapped their sores.

Meanwhile, my mother had arranged a selection of fruit in a wicker basket. She set it on the ground beside the visitors, suggesting they take the basket. They took the fruit but left the basket, and as they disappeared over the ridge, I went to pick it up. "No, Paul!" Mother cried out. "Don't touch it! And don't go near that place where they sat." I watched Father take the basket and burn it, then scrub his hands with hot water and soap. Then Mother bathed my sister and me, though we had had no direct contact with the visitors.

That incident was my first exposure to leprosy, the oldest recorded disease and one of the most dreaded. Although I would have recoiled from the suggestion as a boy of seven, I later felt called to spend my life as a doctor among leprosy patients. I have worked with them almost daily, in the process forming many intimate and lasting friendships among these misunderstood and courageous people.

Over those years, many of the fears and prejudices about leprosy have crumbled, at least in the medical profession. Partly because of effective drugs, it is now viewed as a controllable, barely infectious disease. Leprosy got its fearsome reputation because of the visible damage it does to the body. Even to those who are taking drugs, it remains a disease that can cause severe lesions, blindness, and loss of hands and feet. Only in my lifetime have we learned how this ancient disease produces such terrible effects.

As I studied patients in India, several findings pushed me toward a rather simple theory: perhaps the horrible effects of leprosy come about because leprosy patients have lost the sense of pain. The disease does not spread like flesh-eating bacteria; rather, it attacks a single type of cell, the nerve cell. When that nerve cell falls silent, it no longer warns of danger, and the painless person quite literally destroys his or her own body.

After years of testing and observation, I felt sure that the theory was sound. After many false starts, my team learned to track how the damage occurs. A person uses a hammer with a splintered handle, does not feel the pain, and an infection flares up. Another steps off a curb, spraining an ankle, and, oblivious, keeps walking. Another loses use of the nerve that triggers the eyelid to blink every few seconds for lubricating moisture; the eye dries out and the person becomes blind. Virtually all the devastating effects of leprosy trace back to a single source: one type of nerve cell that has fallen silent.

SOLITARY AND COMMUNAL CELLS

I remember the first time I saw a living cell under a microscope. I had snuck into a college lab early one morning with a teacup of brackish liquid I'd scooped from a pond. Bits of decomposing leaves were floating on top, emitting the musty odor of organic decay.

No sooner had I touched one drop of pond water to a glass slide under the microscope than a universe sprang to life. Hundreds of organisms crowded into view: delicate, single-celled globes of crystal unfurling and flitting sideways, excited by the warmth of my microscope light. I edged the slide a bit, glancing past the more lively organisms. Ah, there it was: an amoeba. A mere chip of translucent blue, barely visible to my naked eye, it revealed its inner workings through the microscope.

This simple, primordial creature performed all the basic functions of my human body. It breathed, digested, excreted, reproduced. In its own peculiar way it even moved, jutting a bit of itself forward and the rest following with a motion as effortless as a drop of oil spreading on a table. After one or two hours of such activity, the grainy, liquid blob would travel a third of an inch. Though in appearance a meager bit of gel, the amoeba manifested *life*, which differs profoundly from mere matter.

That busy, throbbing drop of water gave me a lasting image of the jungle of life and death we share, and beckoned me to further explore living cells.

Years later I am still observing cells, though as a physician I focus now on how they cooperate within the body. I have my own laboratory at a leprosy hospital on swampy ground in the bayou country of Louisiana. Again I enter the lab early, before anyone is stirring, and only the soft buzz of fluorescent lights breaks the quiet.

This morning I will examine a hibernating albino bat who sleeps in a box in my refrigerator. He helps me understand how the body responds to injury and infection. I lift him carefully, lay him on his back, and spread his wings in a cruciform posture. His face is weirdly human, like the shrunken heads in museums. I keep expecting him to open an eye and shriek at me, but he doesn't. He sleeps.

As I place his wing under the microscope lens, again a new universe unfolds. The albinic skin under his wing is so limpid that I can look directly through his skin cells into the vessels underneath. I focus the microscope on one bluish capillary until I can see individual blood cells pushing and thrusting through it. The pulsing fluid is like a river stocked with living matter: a speck of blood the size of this letter *o* contains five million red cells and seven thousand white cells.

I am searching for white blood cells, the body's elite special forces, which protect against invaders. Transparent, bristling with weapons and possessing a Houdini-like ability to slip between other cells, the white cells function as the body's advance guard. Flattened on a microscope slide, they resemble fried eggs sprinkled with pepper.

As I stare, the white cells remind me of the amoeba I first saw as a student in England. Amorphous blobs of liquid, they roam through the bat's body by extending a finger-like projection and hunching along to follow it. Sometimes they creep sideways on the walls of the veins; sometimes they let go and free-float in the bloodstream. To navigate smaller capillaries, the bulky white cells must elongate their shapes, while red blood cells jostle impatiently behind them.

An observer can't help thinking white cells sluggish and ineffective at patrolling territory—until an attack occurs. I take a thin, steel needle and, without waking the bat, prick through its wing to puncture a fine capillary. Instantly, a silent alarm sounds. Muscle cells contract around the damaged capillary wall, damming up the loss of precious blood. Clotting agents halt the flow at the skin's surface. The most dramatic change, though, occurs among the listless white cells.

As if they have a sense of smell, nearby white cells abruptly halt their aimless wandering. Like beagles on the scent of a rabbit, they home in from all directions toward the point of invasion. Their unique shape-changing qualities allow them to ooze between the overlapping cells of capillary walls. When they arrive, the battle begins.

Lennart Nilsson, the Swedish photographer renowned for his remarkable close-ups of activity inside the human body, has captured the battle on film as seen through an electron microscope. In the distance, a shapeless white cell, resembling science fiction's creature "the Blob," lumbers toward a cluster of luminous green bacterial spheres. Like a blanket pulled over a corpse, the cell assumes their shape; for a time the bacteria glow eerily inside the white cell. But those pepper-like dots inside the white cell are granules of chemical explosives, which soon detonate and destroy the invaders. In thirty seconds to a minute, only the bloated white cell remains.

Although the battle often results in the white cell's demise, the death of a single cell has little significance. Besides the fifty billion active white cells prowling the adult human, a backup force one hundred times as large lies in reserve in the bone marrow. When an infection occurs, these reserves leap from the marshes of bone marrow, like beardless young recruits pressed into service. The body can thus mobilize a vast number of white cells; indeed, doctors use a count of them as a diagnostic test to judge the severity of infection.

Each day we live at the mercy of organisms one-trillionth our size. A drop of water may contain as many bacteria as there are humans on earth. Bacteria enshroud my body: when I wash my hands, I sluice five million of them from the folds of my skin. Immunologists share a little joke that they cite when asked how the body can possibly prepare every type of antibody required in our perilous world: GOD, they reply—an acronym for "generator of diversity."

If the body has previously identified a known threat, as in a smallpox vaccination, it imprints certain white cells with a death wish to target that one danger. These cells spend their lives coursing through the bloodstream, alert, scouting. Often the summons to battle never sounds. If it does, however, they hold within them the power to disarm a foreign agent that could destroy every cell in the body.

Medical author Ronald J. Glasser concludes rather humbly, "No matter how we may wish to view ourselves, despite all our fantasies of grandeur and dominion, all our fragile human successes, the real struggle . . . has always been against bacteria and viruses, against adversaries never more than seven microns wide." He describes the process as "a mixture of mystery and chemistry . . . a combination of physics and grace down at the molecular level."

If we as doctors were forced to choose either (1) the human immune system alone or (2) all the resources and technology of science but with the loss of our immune system, without a moment's hesitation we would choose the former. The disease AIDS exposes the helplessness of modern technology when a person's immune system shuts down: pneumonia, cold sores, or even diarrhea can pose a mortal danger.

SPECIALIZATION: LOSS AND GAIN

Dig up a block of forest soil one foot square and one inch deep. According to the naturalist author Annie Dillard, that block contains an average of 1,356 living creatures, including 865 mites, 265 springtails, 22 millipedes, and 19 adult beetles. It would require an electron microscope to reveal the additional trillions of bacteria and the host of fungi and algae.

In a laboratory the scientist begins with our friend the amoeba and works up, classifying from the "lower" to the "higher." What is this term *lower*? How can we trample a million creatures on a hike and return home guiltless? A strict vegan who gulps cold spring water imbibes a horde of creatures—animals!—without flinching. Why do we wince at a bloodied cat along the roadside but take no notice of the billions of tiny animals pulverized by the bulldozer scraping out a roadbed?

The key to our ranking is specialization: the process of cells dividing up labor and limiting their role to a single task. We recognize a more meaningful life in the cat, a higher animal consisting of many cells working together.

The amoeba on my microscope slide occupies the bottom of the zoo-logical ladder. It moves, yes, but no more than a few inches per day. It may spend its lifetime confined to a tin can or the hollow of an old tire.

Unlike humans, it will never tour Europe, visit the Taj Mahal, or climb the Rockies. In order to do that, one needs specialized muscle cells, rows and rows of them, aligned like stalks of wheat. The lower animals skitter, creep, or worm along, covering mere yards of turf. The higher ones hop and leap and gallop or, if winged, vault and soar and dive. Specialization makes the difference.

Consider just one product of specialized cells, the organ of sight. As the husband of an eye surgeon, I hear often about the wonders of the eyes, which take up a mere one percent of the weight of the head. An amoeba has some crude visual awareness: it moves toward light—and that is all. Specialization gives me the ability to gaze through the viewing end of the microscope, noting the subtleties of color in the near-senseless amoeba. The amoeba comprises one cell, whereas I peer at it with 127 million visual cells. Named for their shape, these rods and cones line up in rows to receive images and transmit them to the brain.

Rods, slender and graceful tentacles that extend toward light, out-number the bulbous cones 120 million to 7 million. They are so sensitive that the smallest measurable unit of light, one photon, can excite them; under optimum conditions the human eye can detect a candle at a dis-tance of fifteen miles. Yet with rods alone I would see only shades of gray, as on a moonlit night. Squeezed into the dense forest of rods, the larger cones give me more focal resolution and the ability to distinguish more than a million unique colors.

When it detects a designated wavelength of light, each rod or cone triggers an electrical response to the brain. The brain compiles all these yes-or-no binary messages from rods and cones and—*voilà!*—I get an image of an amoeba swimming on my microscope slide. The feat re-quires so much processing power that half of my brain is devoted directly or indirectly to vision. As I contemplate vision, I am most impressed by this fact: when I see, I remain totally unconscious of cells encoding data and firing, then decoding and reassembling it within the brain. The hos-pital chapel outside my window presents itself not as a series of dots and light flashes, but as a beloved building, whole and meaningful and evoking many fond memories.

Compared to the amoeba's one-celled independence, the lives of my rods and cones may seem dull and stationary. But who among us would trade ends of the eyepiece? For specialization to work, the individual cell must lose all but one or two of its abilities. Although vision cells forgo an amoeba's autonomy and movement, they enable a much "higher," more significant achievement. A single rod or cone can provide me with the wavelength of light that completes my appreciation of a rainbow, a kingfisher plunging into a stream, or a subtle change of expression in the face of a dear friend. Or it may protect me from disaster by firing off a message to the brain when a rock falls from a hillside toward my approaching car.

JOSÉ, REUNITED

An encounter with a patient I'll call José captures for me the importance of membership in the human body and what happens when damaged cells sever the body's connections to the governing brain.

José's body had suffered much damage from leprosy by the time he traveled from Puerto Rico to our leprosy hospital in Louisiana for treatment. By then, research had proved that leprosy does its damage by affecting nerve cells, thus making patients vulnerable to injury. José's insensitivity was so great that, when blindfolded, he could not even detect whether someone was holding his hand. Touch cells and pain cells had fallen silent. As a result, scars and ulcers covered his hands, face, and feet, bearing mute witness to the abuse his body had suffered because it lacked the warning system of pain. Mere stubs on his hands marked where fingers used to be.

Since pain cells in his eyes no longer alerted him when to blink, José's eyes gradually dried out. That condition, aggravated by severe cataracts and glaucoma, soon made him blind. My wife, Margaret, informed him that surgery might correct the cataract problem and restore some vision, but she could not operate until inflammation of the iris went away. Shortly after that, a terrible misfortune cut off José's last link with the outside world. In a last-ditch attempt to arrest the sulfone-resistant leprosy, doctors tried treating him with a new drug, and José had a rare allergic reaction. In a final cruelty, he lost his hearing.

At the age of forty-five, José lost contact with the outside world. He could not see another person, nor hear if someone spoke. Unlike Helen Keller, he could not even use tactile sign language because leprosy had dulled his sense of touch. Even his sense of smell disappeared as the leprosy bacilli invaded the lining of his nose. All his sensory inlets, except taste, were now blocked. Weeks passed and we watched, helpless, as José began to accept the reality of total isolation.

José's body responded with a pathetic mirroring of what was happening to his psyche: his limbs pulled inward toward his trunk and he spent his days curled into a fetal position on the bed. Unable to tell day from night, he would awake from sleep and forget where he was. When he spoke, he did not know if anyone heard or answered. Sometimes he would speak anyway, bellowing because he could not hear the volume, pouring out the inexpressible loneliness of a mind locked in solitary confinement.

In such a state thoughts incurve, stirring up fears and suspicions. José's body coiled tighter and tighter on the bed, preparing for death in the same posture as his birth. Most of us on the staff would pass his room, pause for a moment at the door, shake our heads, and continue walking. What could we do?

Margaret faithfully visited José. Unwilling to watch him self-destruct, she felt she must attempt some kind of radical intervention to restore at least part of his sight. She waited anxiously for the infection in his eye to improve enough for her to schedule surgery.

In order to follow government rules, Margaret faced a nearly insurmountable problem. She must obtain "informed consent" forms for the surgery, but who would sign for José? No one could penetrate through his isolation to ask him for permission. After painstaking research, the hospital staff finally located a sister in Puerto Rico, and the police department there visited her with a surgery release form. The illiterate sister marked an X on a paper, and Margaret scheduled surgery at last, with faint hope of success.

José, of course, did not comprehend what was happening as he was moved to a stretcher and wheeled to the operating room. He lay passive throughout the eye surgery, feeling nothing. After a two-hour procedure, he was bandaged and sent back to his room to recover.

Margaret removed the bandages a few days later, an experience she will never forget. Although José had sensed some gross movement and had probably reasoned someone was trying to help him, nothing prepared him for the result. He got the use of one eye back and could see again. As his eye struggled against the bright light and slowly brought into focus the medical people gathered around the bed, the face that had not smiled in months cracked into a huge, toothless grin.

During that time of solitude, José's brain had floated intact inside his skull, complete with memory, emotions, and instructions for directing his body. Suddenly human contact was restored. José made it known that he wanted his wheelchair parked at the door to his room all day long. He would sit there quietly, every few seconds glancing up and down the long corridors of the leprosarium. When he saw another person approaching, his face would break into that irrepressible smile.

José insists on coming to our small church every Sunday, even though he can hear nothing of the service. With stubby fingers, he can barely grasp the control knob of his electric wheelchair, and his narrow tunnel vision causes him to bump into objects up and down the hospital corridors. Other attenders have learned to greet him by stooping down, putting their faces directly in front of his, and waving. José's wonderful smile breaks out, and sometimes his bellowing laugh. Although he cannot see well, and still cannot hear or feel, somehow he can sense the fellowship of the church. He has rejoined the community, and for him that is enough.

CHAPTER FOUR

DIVERSITY

The Richness of Life

IN MY MEDICAL LABORATORY, one drawer contains neatly filed specimens of an array of cells from the adult human body. Lifeless, excised from the body, stained with dyes and mounted in epoxy, they don't do justice to the churn of active cells inside me at this moment. Even so, when I parade them under the microscope some impressions about the body take shape.

The cells' diversity stands out first. Though they share a chemical makeup, the body's cells are as different from each other as the animals in a zoo. Red blood cells, discs resembling Life Savers candies, have voyaged through my blood vessels bearing oxygen supplies for the other cells. The muscle cells that absorbed so much of that nourishment stretch out sleekly and supplely. Cartilage cells with shiny black nuclei look like bunches of black-eyed peas glued together.

Fat cells seem lazy and leaden, resembling overstuffed white, plastic garbage bags. Cut in a cross section, bone brings to mind the rings of a tree, its cells overlapping to provide strength and solidity. In contrast, skin cells arrange themselves in undulating patterns of softness and texture, giving shape and beauty to our bodies. They curve and jut at unpredictable angles so that every person has a unique fingerprint—let alone a unique face.

The aristocrats of the cellular world are dedicated to reproduction. A woman's contribution, the egg, is one of the largest cells in the body, its ovoid shape just visible to the unaided eye. It seems fitting that all other cells in the body should derive from this simple, primordial design. Offsetting the egg's quiet repose, the male's sperm cells, tiny tadpoles with distended heads and skinny tails, compete for position as if aware that only one of billions will gain the honor of fertilization.

The nerve cell, the one I have devoted much time to studying, has about it an aura of wisdom and complexity. Web-like, it extends to unite the body's parts with an electrical network of dazzling sophistication. Its axons, organic "wires" that carry far-flung messages to and from the human brain, can reach a yard in length.

I never tire of viewing these oddly varied specimens. Individually they seem humble, yet I know these hidden parts combine their efforts to give me the richness of life. Every second of every day my smooth muscle cells modulate the width of my blood vessels, gently push waste matter through my intestines, open and close the plumbing in my kidneys. When things are going well—my heart contracting rhythmically, my brain humming with input, my lymph bathing tired cells—I rarely give my loyal cells a passing thought.

THE SPICE OF LIFE

I believe these diverse cells in my body can also teach me about larger organisms: families, groups, communities, villages, nations—and especially about the community that is likened to a body more than thirty times in the New Testament. I speak of that network of people scattered across the planet who have little in common other than their membership in the Body that follows Jesus Christ.

It seems safe to assume that God enjoys variety, and not merely at the cellular level. Not content with a thousand insect species, the Creator conjured up several hundred thousand species of beetles alone. In the famous speech at the end of the book of Job, God points with pride to such oddities of creation as the mountain goat, the wild ass, the ostrich, and the crocodile. The human species, made in God's image, includes pygmies and Nubians, pale Scandinavians and swarthy Egyptians, big-boned Russians and petite Japanese.

Humans have continued the diversification, grouping themselves according to distinct cultures. Consider the continent of Asia. In some countries women wear long pants and men wear skirts. In tropical Asia people drink hot tea and munch on blistering peppers to keep cool. Japanese prefer their ice cream fried and their fish raw. Westerners puzzle over the common Indian custom of marriages arranged by parents; many Indians gasp at anyone entrusting such a decision to fickle romantic love. And when the British introduced the violin to India a century ago, men started playing it while sitting on the floor, holding it between the shoulder and the sole of the foot. Why not?

On my international travels I am struck by the world's incredible diversity, and churches reflect that cultural self-expression. For too long they mimicked Western ways so that hymns, dress, architecture, and church names were the same around the world. Now indigenous churches are adapting their own expressions of worship. I must guard against picturing the spiritual Body as comprising only American or British cells; it is far grander and more luxuriant.

African Americans in the Southern United States shout their praises to God. Believers in Austria intone them, accompanied by showpiece organs and illuminated by stained glass. Some Africans dance their praise, following the beat of a skilled drummer. Sedate Japanese Christians express their gratitude by creating objects of beauty. Indians point their hands upward, palms together, in the *namaste* greeting of respect, which has its origin in the Hindu concept, "I worship the god I see in you," but gains new meaning as Christians use it to recognize the image of God in others.

The church of Christ, like our own bodies, consists of individual, unlike cells that are knit together to form one Body.

A MOTLEY CREW

Just as my body employs a bewildering zoo of cells, no one of which resembles the larger body, so the spiritual Body comprises an unlikely assortment of humans. *Unlikely* is the right word, for we are decidedly unlike one another and the One we follow.

The first humans disobeyed the only command God gave them. Abraham, the leader God chose to head a new nation known as "God's people," tried to pawn off his wife, Sarah, on an unsuspecting Pharaoh. Sarah herself, when informed at the ripe old age of ninety-one that God was ready to deliver the promised son, broke into rasping laughter. The harlot Rahab became revered for her great faith. And Solomon, the wisest man who ever lived, proceeded to flout every proverb he had so astutely composed.

After Jesus, the pattern continued. The two disciples who did most to spread the word after his departure, John and Peter, were the two he had rebuked most often for petty squabbling. The apostle Paul, who wrote more books than any other Bible writer, was chosen while kicking up dust whirls in search of Christians to torture. Jesus had nerve, entrusting the high-minded ideals of love and unity and fellowship to this group.

Little wonder cynics have looked at the church and sighed, "If that group of people is supposed to represent God, I'll decide against God." Or, as Nietzsche put it, "His disciples will have to look more saved if I am to believe in their savior."

The church contains a collection of people as diverse as the cells in the human body. I think of the churches I have known: Is there another institution encompassing such a human mosaic? Young idealists wearing T-shirts and sporting tattoos share the pews with executives in suits. Bored teenagers tune out the sermon even as their eager grand-parents turn up their hearing aids. Some members gather as methodi-cally as a school of fish, then quickly disperse to return to their jobs and homes. Others migrate together like social amoebae and form intentional communities.

During my life as a missionary surgeon in India and now as a member of the small chapel on the grounds of the leprosy hospital in Louisiana, I have seen my share of unlikely seekers after God. I must admit that most of my worship has taken place among people who do not share my tastes in music, sermons, or even thought. Still, over those years I have been profoundly—and humbly—impressed that I find God in the faces of my fellow worshipers, people who are shockingly different from each other and from me.

C. S. Lewis recounts that when he first started going to church, he disliked the hymns, which he considered to be fifth-rate poems set to sixth-rate music. Then, he writes, "I realized that the hymns (which were just sixth-rate music) were, nevertheless, being sung with devotion and benefit by an old saint in elastic-side boots in the opposite pew, and then you realize that you aren't fit to clean those boots. It gets you out of your solitary conceit."

A color on a canvas can be beautiful in itself. The artist excels, however, not by slathering a single color across the canvas but by positioning it between contrasting or complementary hues, so that the original color derives richness and depth from its surroundings.

The basis for unity within any human community begins not with our similarity but with our diversity.

BODILY STATUS

In a leprosy patient, the millions of healthy cells in a hand or foot, or the watchful rod and cone cells in the eye, can be rendered useless due to the breakdown of a few nerve cells. Similarly, in sickle cell anemia or leukemia, the malfunction of a single type of cell can result in death. And if the cells assigned to keep kidney filters in repair fail, a person may soon die of toxic poisoning. These targeted diseases prove that the body needs *each* of its many members in order to thrive.

In medical school I learned about crucial cells that make their entrance for one dramatic act, then disappear. Before birth, only a third of the fetus's blood—the amount needed to nourish developing lung tissue—travels to the dormant lungs, since the fetus receives its oxygen through the placenta. A special blood vessel, the *ductus arteriosus,* shunts most of the blood to the rest of the body. Suddenly, at the very moment of birth, all the blood must take a new route through the lungs for oxygenation. The midwife or doctor waits anxiously for the baby to take its first breath.

To accomplish this change, an amazing event occurs. A flap descends like a curtain, deflecting the blood flow back to the aorta. Over the next few days a customized muscle squeezes shut the *ductus arteriosus.* The muscle exists only for this essential act. If it fails to perform its designated task, the baby may die, apart from surgical intervention. If it succeeds, the heart permanently seals the *ductus arteriosus* and the body gradually absorbs it. On this little-known group of transitory cells, every human life depends.

Appropriately, the Bible stresses this very quality, the worth of every member, in its imagery of the body. Listen to the mischievous way in which the apostle Paul presents the analogy in 1 Corinthians 12:

> Those parts of the body that seem to be weaker are indispensable, and the parts that we think are less honorable we treat with special honor. And the parts that are unpresentable are treated with special modesty, while our presentable parts need no special treatment. But God has put the body together, giving greater honor to the parts that lacked it, so that there should be no division in the body, but that its parts should have equal concern for each other. If one part suffers, every part suffers with it; if one part is honored, every part rejoices with it.

Paul's point is clear: the body needs every single member for its proper health and function. More, the less visible, "unpresentable" members—I think of organs like the pancreas, kidney, liver, and colon—may be the most valuable of all. Although I seldom feel consciously grateful for them, they perform vital tasks that keep me alive.

I need this reminder because human societies tend to assign worth based on a hierarchy of value. For example, airlines reward highly trained pilots with fine salaries and fringe benefits. Within the corporate world, such symbols as titles, office size, and stock options signal the worth of any given employee. In the military, a sergeant salutes superior officers and gives orders to those of lower rank; the uniform and stripes alert everyone to the soldier's relative status.

Living in such a society, my vision gets clouded. I begin viewing janitors as having less personal worth than software developers. When that happens, I must turn back to the lesson from the body, which the apostle Paul spells out. Human society confers little status on janitors because their position is considered unskilled. The Body, however, recognizes that lowly janitor cells are indispensable to overall health. If you doubt that, ask someone who must go in for kidney dialysis three times per week.

In my own field of medicine, I marvel at the valuable contributions of nursing aides, orderlies, and nurses, whose salaries are much lower than the physicians' and administrators'. Oliver Sacks writes of a time when these lesser-paid staff went on strike at his hospital and he organized medical students to fill in for the sake of the patients:

> We spent the next four hours turning patients, arranging their joints, and taking care of their toilet needs, at which point the two students were relieved by another pair of students. It was backbreaking, round-the-clock work, and it made us realize how hard the nurses and aides worked in their normal routine, but we managed to prevent skin breakdown or any other problems among the more than five hundred patients.

The Bible directs harsh words to those who show favoritism. The book of James spells out a situation we can all identify with:

> Suppose a man comes into your meeting wearing a gold ring and fine clothes, and a poor man in filthy old clothes also comes in. If you show special attention to the man wearing fine clothes and say, "Here's a good seat for you," but say to the poor man, "You stand there," or, "Sit on the floor by my feet," have you not discriminated among yourselves and become judges with evil thoughts? (James 2:2-4)

In a society that ranks everything from football teams to "the best chili in New York," an attitude of relative worth can easily seep into the church. The association of people who follow Jesus, though, should not operate like a military machine or a corporation. The church Jesus founded is more like a family in which the special-needs child has as much worth as his sister the Rhodes scholar. It is like the human body, composed of cells most striking in their diversity but most effective in their mutuality.

If every cell accepts the needs of the whole Body as their purpose, then the Body will live in health. It is a brilliant stroke, the only pure egalitarianism I observe in all of society. God has endowed every person in the Body with the same capacity to respond. A teacher of three-year-olds has the same value as a bishop, and in the end, that teacher's work may prove just as significant. In God's eyes, a widow's dollar can match a millionaire's bequest. Shyness, beauty, eloquence, race, sophistication—none of these matter, only loyalty to the Head, and through the Head to one another.

LOU'S ONE GIFT

Our little church at Carville, Louisiana, includes a devout Christian named Lou, Hawaiian by birth, whose body manifests the ravages of leprosy. Lacking eyebrows and eyelashes, Lou's face seems unbalanced, and paralyzed eyelids cause tears to overflow as though he is constantly crying. He has become almost totally blind because of the failure of nerve cells on the surface of his eyes.

Like many other leprosy patients, Lou struggles with a growing sense of isolation. His sense of touch has faded now, and that, combined with his near blindness, makes him afraid and withdrawn. He most fears that his sense of hearing may also leave him, for Lou's main love in life is music. He can contribute only one gift to our church, other than his physical presence: he sings hymns to God while accompanying himself on an autoharp. Our physical therapists designed a glove that permits Lou to play the instrument without damaging his insensitive hand.

But here is the truth about the Body: no person in Carville contributes more to the spiritual life of our church than Lou playing his autoharp. He affects us all by offering as praise to God the limited, frail tribute of his music.

When Lou departs, he will create a void in our church that no one else can fill—not even a professional harpist with nimble fingers and a degree from the Juilliard School. Everyone in the church knows that Lou contributes as a vital member, no less important than any other member—and therein lies the secret of Christ's Body. If each of us can learn to glory in the fact that we matter little except in relation to the whole, and if each will acknowledge the worth in every other member, then perhaps the cells of Christ's Body will begin acting as Christ intended.

CHAPTER FIVE

UNITY

The Sense of Belonging

PICTURE A BIOLOGIST removing from the incubator an egg containing a nearly developed young chicken. Fourteen days ago this egg consisted of a single cell (the world's largest single cell is an unfertilized ostrich egg). Now it has divided into hundreds of millions of cells, a whirlpool of protoplasm rearranging itself to prepare for life outside. The biologist cracks the shell and sacrifices the chick.

Word travels fast through the body, but hours may pass before the far outposts surrender their hold on life. From the tiny heart the biologist extracts a few muscle cells, still living though in a dead embryo, and drops them in saline solution. Under the microscope they appear as long, spindly cylinders, crisscrossed like sections of railroad track. Their destiny is to throb, and they persist even in the anarchic world beyond the body. Now isolated from the chick, each cell continues its pitiful and useless palpitations.

Ungoverned by a pacemaker, the cardiac cells beat spasmodically, in a rhythm approximate to the chick's normal 350 beats per minute. As the observer watches over a period of hours, however, something marvelous takes place. Instead of five independent heart cells contracting at their own pace, at first two, then three, and then all the cells pulse together in unison. Five beats converge into one. How do cells communicate the urge for unity, and why?

Some species of fireflies show a similar pattern. A wanderer discovers a tribe of them flickering haphazardly in a forest clearing. As she watches, one by one the fireflies synchronize until soon she sees not dozens of twinkling lights but one light, switched on and off, with fifty branch locations. Heart cells and fireflies both sense an innate rightness about playing the same note at the same time.

Wherever we look, it seems, community is the order of the day. At the smallest observable levels cooperation prevails, and we could neither breathe nor eat without it. Producing the oxygen by which we live requires colonies of bacteria to aid plant photosynthesis, and our digestion relies on similar colonies to help break down what we eat. Recent studies have determined that the human body contains about as many bacteria, thirty-nine trillion, as the person's "own" cells. We encompass an entire ecosystem, called the human microbiome.

Hosts and guests alike must cooperate together to produce a functioning person. What mysterious force unites the cells in my body so that they all act like Paul Brand (with a few rebellious exceptions)?

AN URGE TO BELONG

Unity is the foundation of bodily life, where every heart cell obeys in tempo or the animal dies. How does the roaming white blood cell in the bat's wing know which cells to attack as invaders and which to welcome as friends? No one yet knows, but the body's cells have a near infallible sense of belonging.

The body senses infinitesimal differences with an unfailing scent. My body knows which cells belong to Paul Brand and maintains constant vigilance against intruders. The first transplant recipients did not die because their new kidneys failed, but rather because their bodies would not be fooled. Though the new kidney cells looked and acted in every respect like the old ones, they did not belong. Transplant surgeons must now give the recipient immunosuppressant drugs for the rest of the patient's life in order to lull the guards and keep them from sounding an alarm against the transplanted organ.

To complicate the process of identity, the composite of Paul Brand today—bone cells, fat cells, blood cells, muscle cells—differs almost entirely from my components two decades ago. All cells have turned over their mission to fresh recruits (except for nerve cells and brain cells, most of which do not get replaced). More like a fountain than a sculpture, my body maintains its continuity while constantly being renewed. Somehow my body knows the new cells belong and welcomes them.

Occasionally children are born without an immune system, the "bubble boy" syndrome. Until recent advances in therapy, they had to spend their lives in a plastic tent, untouched by other humans. NASA rigged up a bulky spacesuit for one such child, who then tugged behind him a golf cart-size contraption that scrubbed the air of impurities. Lacking a sense of shared identity, this unfortunate boy's cells welcomed all intruders, including lethal bacteria and viruses.

The secret to membership lies locked away inside each cell nucleus, chemically coiled in a strand of DNA. The three billion letters of DNA spell out instructions that, if printed in a tiny font, would fill three hundred books of a thousand pages each. (Each letter counts: a mistake in only two letters can cause a disease like cystic fibrosis.) A nerve cell may operate according to instructions from volume four and a kidney cell from volume twenty-five, yet each carries the whole compendium, its credential of membership in the body. The entire body could be reassembled from information in any one of the body's cells—which forms the basis for cloning and for the evolving technology of stem cell transplants.

As a Christian, I believe that the Designer of DNA further challenged the human race to a new and higher purpose: membership in a spiritual Body. The community that the New Testament calls the Body of Christ differs from every other human group: unlike a social or political body, joining it requires an identity transfer, analogous to an infusion of DNA. Jesus described the process to Nicodemus as being "born again" or "born from above," indicating that spiritual life requires an identity change as drastic as a person's first entrance into the world. We become, quite literally, God's children.

"By him we cry, 'Abba, Father.' The Spirit himself testifies with our spirit that we are God's children," wrote the apostle Paul (Romans 8:15-16). We are "in him" and he is "in us," the New Testament says in several places. We members take on Christ's name and identity, and he asks from us the same kind of loyalty and unity that my own body's cells give to me.

That common identity links all members of Christ's Body with a unifying bond. I sense that bond when I meet strangers in India or Africa or California who share my loyalty to the Head; instantly we become brothers and sisters, fellow cells in Christ's Body.

In a healthy church, unity trumps diversity. Paul, who as a faithful Pharisee used to thank God every day that he was not born a slave, a Gentile, or a woman, changed dramatically. "There is neither Jew nor Gentile, neither slave nor free, nor is there male and female, for you are all one in Christ Jesus," he told the fractious Galatians (3:28). Such ethnic and gender categories melt away in significance compared to the new identity that we share.

The process of joining Christ's Body may at first seem like a renunciation. I forfeit autonomy. Ironically, however, renouncing my old value system—in which I had to compete with other people on the basis of power, wealth, and talent—and committing myself to the Head abruptly frees me. My sense of competition fades. No longer do I have to compete through life, looking for ways to prove myself. Instead, I have the singular goal of pleasing God, of living for an audience of One. More, I can partner with other cells in the Body to accomplish God's work in the world.

HOMEOSTASIS

Driving home one evening during a heavy downpour, I suddenly see a small, dark shape scurrying out onto the road—probably an armadillo or opossum. Before that thought even registers, my foot has instinctively tapped the brake pedal.

I feel the sickening, out-of-control sensation of a skid coming on as the rear of the car hydroplanes off to the right. My hands grip the steering wheel more tightly. In response to a few quick jerks of my wrist, the car fishtails and finally straightens out. Once I have steering back under control, I breathe deeply and slow down until my anxiety subsides.

The entire incident lasts maybe three seconds. When I arrive home, I will recount to my wife what happened: an animal crossed a rain-slicked highway, and I arrested a dangerous skid just in time. Those are the external events, simple and matter-of-fact. The rest of the way home, still keyed up from the adrenaline pumping through my body, I think back on a few of the internal events.

Few parts of my body went untouched by the momentary crisis. My brain relied on a reflex response to direct my foot onto the brake pedal. At the same time, my hypothalamus ordered up chemicals that, with lightning speed, equipped me to cope.

Vision intensified as my pupils dilated and my eyes widened to admit more light and a larger visual field. My heart beat faster and contracted more forcefully, even as vascular muscles relaxed in order to allow blood vessels to widen for increased blood flow. My muscles went on alert. The makeup of my blood changed: more blood sugars surged in to provide emergency reserves for those muscles, and clotting materials multiplied in preparation for wound repair. Bronchial tubes in my lungs flared open to allow a faster oxygen transfer.

On the skin, blood vessels contracted, bringing on a pale complexion ("white as a ghost"); the reduced blood flow lowered the danger of surface bleeding in case of injury and freed up more blood for the muscles. The electrical resistance of skin changed as a protective response against potential bacterial invaders. Sweat glands activated to increase the traction of my palms on the steering wheel.

Meanwhile, nonessential functions slowed down. Digestion nearly came to a halt—blood assigned to that and to kidney filtration was redeployed for more urgent needs.

Fear, relief, heightened awareness—I felt all these sensations, and for the next twenty miles or so they made me a better driver. Yet inside my body a full-scale campaign had been launched to equip me for the classic alternatives of fight or flight. And what skilled executive coordinated the different responses of trillions of cells? A single chemical messenger called adrenaline.

We experience the effects of adrenaline every day: a clap of thunder startles us, we hear a bit of shocking news, we drive through a dangerous neighborhood, we stumble and nearly fall. Adrenal reactions occur so smoothly and synchronously that we rarely, if ever, stop to reflect on all the elements involved. Yet adrenaline is just one of many hormones at work in my body coaxing a cooperative response from diverse cells.

Medicine has coined a wonderful word to describe how the body unites its many cells to serve the whole: homeostasis. A physician and writer named Dr. Walter Cannon introduced the term in his classic study *The Wisdom of the Body* and also coined the term "fight-or-flight response." He viewed the body as a community that consciously seeks out the most favorable conditions for itself. It corrects imbalances in fluids and salts, mobilizes to heal itself, and deploys resources on demand, all with the goal of maintaining a dependable *milieu interieur,* as the French call it.

You can see homeostasis on vivid display in modern hospitals, where monitors record a patient's pulse and other vital functions. I visit a patient in a hospital room. As I enter, red numerals glow a steady 70, her resting pulse. She notices my presence and greets me, the flurry of emotional responses shooting her pulse up to 91. She reaches over to shake my hand, and the rate surges past 100. During my visit, the numbers rise and fall in concert with her moods and actions. A sneeze causes the most violent reaction of all, a pulse of 110.

Cells are constantly calling out their needs, and the body responds to each demand. Kidneys adjust to the body's needs, increasing or decreasing the amount of fluid and minerals eliminated. After unusual exertion, they may stop the outflow altogether to prevent dehydration; hence a triathlete may not urinate for twenty-four hours after a race.

Sweat. I could write an entire chapter on that model of homeostasis. What a lizard wouldn't give for warm blood and sweat glands! On brisk mornings the reptile must wriggle over to the sunlight and warm up before it can start climbing trees and catching flies. If the lizard overheats, it frantically scuttles toward shade. In humans, however, an efficient cooling system uses sweat to cool our bodies to a constant internal temperature so that sensitive organs can maintain a *milieu interieur*. Otherwise, we could hardly function in a climate where temperatures exceed eighty degrees Fahrenheit.

The Japanese physiologist Yas Kuno spent thirty years studying sweat, and in 1956 he published a 416-page book, *Human Perspiration*. He found the body to be so sensitive that a change of one-tenth of a degree in temperature sets off alarms in the skin's thermoreceptors. Humans have the finest cooling system of all mammals; most animals will run fevers on a hot day. A marathon runner may shed three to five quarts of fluid in a three-hour race, but inside his temperature will hardly waver. (Animals compensate, though. A dog or tiger pants, creating its own internal fan. An elephant finds a water hole and wades in for a hose-down.)

All these operations—heart rate, fluid control, perspiration—adapt second by second as the body seeks the very best state. Hormone-like compounds, the prostaglandins, bathe the body's cells: one lowers blood pressure and another raises it; one initiates inflammation, another inhibits it. These messenger fluids travel from cell to cell, visiting nearly every tissue of the body, linking isolated cells and organs into units of a coordinated response.

Until recently, anatomists believed that glands such as the adrenal and pituitary sent out their hormonal instructions independently. New discoveries point to reliance on the brain at virtually every point. Instructions on growth, on deployment of resources, and on how to meet a crisis all originate in the head, which senses the needs of the entire body.

In the human body, the sense of belonging extends two ways: a cell follows orders from the brain, while also recognizing a bond with every other cell in the body. So too in the spiritual Body. God calls me into an organic community, and I join a Body that binds me to other diverse cells. "We will grow to become in every respect the mature body of him who is the head, that is, Christ. From him the whole body, joined and held together by every supporting ligament, grows and builds itself up in love, as each part does its work" (Ephesians 4:15-16).

The word *another* hints at a kind of cell-to-cell cooperation, and we cannot escape the word in the New Testament. It appears as a relentless leitmotif. "Accept one another," we are told. "Serve one another" by "washing one another's feet." We are to confess our sins to one another, pray for one another, forgive one another, teach and admonish one another, comfort one another, and bear one another's burdens.

Jesus left us the most inclusive command of all: "Love one another as I have loved you." When we obey the Head, following the orders that coordinate the Body's many parts, unity results.

AN ACT OF ONENESS

Human society approaches such unity only rarely. Families achieve it sometimes, as in the powerful tug of loyalty that binds me to my children scattered around the world. During a crisis, such as an earthquake or a forest fire, a town or even an entire nation may join together in common cause.

Jesus prayed for an even richer experience of unity in his Body. He asked "that all of them may be one, Father, just as you are in me and I am in you. May they also be in us so that the world may believe that you have sent me" (John 17:21). Do we in the church catch the vision of that unity—a unity based not on social class or interest group or kinship or race, but on common belonging in Jesus Christ?

Sadly, we see many examples of *dis*unity in the church. I have, though, seen what can happen when the Body truly welcomes a new member. Those scenes give me a lasting vision of God at work in the world. I will mention only one example.

John Karmegan came to me in Vellore, India, as a leprosy patient in an advanced state of the disease. We could do little for him surgically since his feet and hands had already been damaged irreparably. We could, however, offer him a job and a place to stay.

Because of one-sided facial paralysis, John could not smile normally. When he tried, the uneven distortion of his features would draw attention to his paralysis. People often responded with a gasp or a gesture of fear, and so he learned not to smile. Margaret, my wife, stitched his eyelids partly closed to protect his sight. Though grateful for her efforts, John grew more and more paranoid about what others thought of him.

Perhaps in reaction to his marred appearance, John acted out the part of a troublemaker. I remember many tense scenes in which we had to confront him with some evidence of stealing or dishonesty. He treated fellow patients cruelly and resisted authority, sometimes organizing hunger strikes against the leprosy hospital. By almost anyone's reckoning, he was beyond rehabilitation.

Perhaps John's very irredeemableness attracted my aging mother to him, for she often latched onto the least attractive specimens of humanity. She spent time with John and eventually led him into the Christian faith. He was baptized in a cement tank on the grounds of the leprosarium.

Conversion, however, did not temper John's high dudgeon against the world. He gained some friends among fellow patients, but a lifetime of rejection and mistreatment had permanently embittered him against all nonpatients. One day, almost defiantly, he asked me what would happen if he visited the local Tamil-speaking church in Vellore.

I went to the leaders of the church, described John, and assured them that despite obvious deformities, he had entered a safe phase of the arrested disease and would not endanger the congregation. They agreed he could visit. "Can he take Communion?" I asked, knowing that the church used a common cup. They looked at each other, thought for a moment, and agreed he could also take Communion.

Shortly thereafter I took John to the church, which met in a plain, whitewashed brick building with a corrugated iron roof. I could hardly imagine the trauma and paranoia inside a leprosy patient who attempts for the first time to enter that kind of setting. As I stood with him at the back of the church, his paralyzed face showed no reaction, but his body's slight trembling betrayed his inner turmoil. I prayed silently that no church member would show the slightest hint of rejection.

As we entered during the singing of the first hymn, an Indian man toward the back of the church turned and saw us. We must have made an odd couple: a white person standing next to a leprosy patient with patches of his skin in garish disarray. I held my breath.

And then it happened. The man put down his hymnal, smiled broadly, and patted the chair next to him, inviting John to join him. John could not have been more startled. Haltingly, he made shuffling half-steps to the row and took his seat. I breathed a prayer of thanks.

That one incident proved to be the turning point of John's life. Years later I visited Vellore and made a side trip to a factory that had been set up to employ disabled people. The manager wanted to show me a machine that produced tiny screws for typewriter parts. As we walked through the noisy plant, he shouted at me that he would introduce me to his prize employee, a man who had just won the parent corporation's all-India prize for the highest quality work with fewest rejects. As we arrived at his work station, the employee turned to greet us, and I saw the unmistakable crooked face of John Karmegan. He wiped the grease off his stumpy hand and grinned with the ugliest, loveliest, and most radiant smile I have ever seen. He held out for my inspection a palmful of the small precision screws that had won him the prize.

A simple gesture of acceptance may not seem like much, but for John Karmegan it proved decisive. After a lifetime of being judged on his damaged appearance, he had finally been welcomed on a different basis. God's Spirit had prompted the Body on earth to adopt a new member, and at last John knew he belonged.

CHAPTER SIX

THE ECSTASY OF COMMUNITY

I AM SITTING IN MY CLUTTERED OFFICE on a lazy summer day, leaning back in my chair. In the spirit of mindfulness, I decide to concentrate on reports from my sense organs, starting with my eyes.

Around me, stacks of journals, notes for books in process, and unanswered correspondence collect in ragged, top-heavy piles. They oppress me, so I pad over to a window. I glance at my vegetable garden, and a pang of guilt reminds me I have not watered and fertilized it recently. Just to the right, however, the plant that gives me greatest delight, the fig tree, is bearing fruit in full glory.

Pendulous figs in velvety shades ranging from green to purple dangle so thickly off every branch that the entire tree bows. Each year when the figs ripen, a population of admiring butterflies suddenly appears, and thousands of them now encircle my fig tree in a shifting corona of color. I can actually hear the papery sound of their beating wings. I watch as the butterflies test each tempting fruit with a "tongue" smaller in diameter than a thread. They light on the unripe figs momentarily, linger a few seconds at those just turning red, and settle in to gorge themselves on the figs two days past perfect ripeness. I have learned a foolproof method of selecting perfect figs: pick the ones that butterflies loiter on but do not pierce.

Sounds reach my ears: my mongrel dog snuffling around in a corner, the deep throb of a barge on the Mississippi River, the distant chatter of a lawnmower, piano music wafting in from my daughter's practice room. The lawnmower gives rise to the pungent aroma of cut grass. If I tilt my head a bit and sniff, I can also smell the sweet fermentation of figs on the ground. Both these scents are partly spoiled by a more pervasive, sulfurous odor from the petrochemical factory down the river.

On one level, nothing much is happening today. By attending to my environment, however, I realize very much is happening. My nose, eyes, and ears had been recording all those sensations even before I consciously tuned in to them. So important in forming my view of the world, these senses merit closer scrutiny.

HEARING

"God gave man two ears," remarked Epictetus the Stoic, "but only one mouth, that he might hear twice as much as he speaks." Compared to those of an elephant or a rabbit, human ears seem puny and underdeveloped. They capture far less sound than a dog's or horse's ears, and cannot compete for ear expressiveness—we wiggle ours only as a party trick. Even so, the organs of hearing serve us well. The pliable eardrum can register sounds as soft as the drop of a straight pin and as noisy as a New York subway, one hundred trillion times louder.

High school biology students learn what happens after the eardrum vibrates: three miniature bones, informally known as the hammer, anvil, and stirrup, transfer that vibration into the middle ear. As an orthopedist, I have worked with most of the bones in the human body, and none impress me more than this trio, the body's smallest. Unlike every other bone, these do not grow with age; a one-day-old infant has a fully developed set. They are in perpetual motion since every perceptible sound causes these tiny bones to swing into action.

How do I distinguish two different sounds, such as the buzz of a fly droning about in my room and the rumble of the lawnmower a block away? Every distinct sound has a signature of vibrations per second. If your ears detect a wave of molecules oscillating 256 times per second, for example, you are hearing the musical middle C. A tuning fork demonstrates the process, for its tines visibly move back and forth when struck.

Inside an inch-long chamber known as the organ of Corti, twenty-five thousand sound-receptor cells line up to receive these vibrations, like strings of a piano waiting to be struck. A few of these cells will fire off signals to the brain when a 256-cycle vibration reaches them, and I thus recognize a middle C. The others await their own programmed frequency. Imagine the bedlam of cellular activity when I sit before a full orchestra and hear twelve notes at once, as well as the variety of musical textures from many different instruments.

Except in the case of extremely loud sounds, the vibration itself never reaches the brain. Instead, the transmission process resembles the digital coding on a compact disc or MP3 player. The brain receives messages from sound receptors in a series of on-or-off blips, sorts them out, and pieces together the meaningful result.

Of course, the brain makes its own contributions. I experienced this in a most poignant way when my wife and I celebrated our fortieth wedding anniversary. The phone rang, and Margaret and I picked up extensions simultaneously. "Hello, Mom. Hello, Dad. Congratulations!" we heard, recognizing the voice of our son, Christopher, in Singapore. Then, to our surprise, we heard the same words again, this time from our daughter Jean in England. And then again from Mary in Minnesota, Estelle in Hawaii, Patricia in Seattle, and Pauline in London. Our six children had conspired to place a globe-girdling conference call.

Those sounds transported me back to scenes around the family dinner table when we laughed and teased together. The voices of my six children instantly brought tears to my eyes and filled me with joy. All the warmth of my love for them and the history of our shared lives surged up at once. The sounds, which began as mechanical forces from thousands of miles away, touched the person inside the computer brain, the "ghost in the machine" in Gilbert Ryle's term.

The brain even has the ability to simulate sound when there is none. If I let my mind drift even now, I can hear the four crashing chords of Beethoven's Fifth Symphony, the melodious voice of my daughter Pauline, the piercing tones of a London air raid siren. Apart from any vibration of molecules and firing of receptor cells, somehow my brain resurrects the sounds from stored memory.

SMELL AND TASTE

I write of hearing with a sense of wonder but of smell with near incredulity. Certain phrases recur in textbooks describing smell: "difficult to explain," "not yet determined," "it is still not understood precisely how."

A male moth comes across a single molecule of a pheromone emitted by a female three miles away. He will not eat or rest until he finds the one who tantalizes him, and one molecule per mile will suffice to track her down. Or consider a salmon that leaves a river in Oregon as a mere fingerling and voyages far into the ocean, thousands of miles from home. Without a map or visual signposts, with no clues other than its sense of smell, the adult salmon will find its way back to the stream of its birth.

Smell compels action. A pig will excavate earth like a bulldozer in pursuit of a truffle; a bear will rip down a tree branch and brave a hundred stings for a slurp of honey. A male boll weevil will passionately attempt to mate bolls of soft cotton all day long when the fields are sprayed with a female's scent.

What humans lack in smell intensity, we make up for in variety. We have between six and ten million neurons devoted to smell, each one relying on olfactory receptors devoted to specific types of chemicals. The signals from overlapping receptors allow us to distinguish an enormous spectrum of smells.

Taste deserves mention, of course, as one of the five major senses. "Gastronomy rules all life," wrote the nineteenth-century, French epicure Jean Anthelme Brillat-Savarin. "The newborn baby's tears demand the nurse's breast, and the dying man receives, with some pleasure, the last cooling drink."

The experience of taste stimulates gastric juices in the same way the smell of sizzling steak or frying bacon can awaken in us a sudden, unexpected hunger. If a hospital patient "primes" food by tasting it first, before having the food administered through a feeding tube or intravenously, the body will absorb more nourishment. Taste also serves as a barrier to keep us from putting poisons and toxic substances into our mouths, many of which we instinctively reject as bitter.

Even so, it takes far more of a substance to stimulate a stubby taste bud than it does to register on a smell receptor. Taste, in fact, relies mostly on smell, as any gourmand with a stopped-up nose can confirm. The two senses together played a leading role in human history. Absent the craving for spices that led to expeditions, the Americas might have lain "undiscovered" by Europeans for another century.

The amount of substance needed to trigger smell defies belief. No laboratory can perform an analysis with a hundredth the speed and accuracy of a bloodhound's nose. A detective holds a sock before a baleful dog, who has forty times more olfactory cells than any human. The bloodhound sniffs deeply a few times, sorting out stale cigarette smoke, the artificial odor of Dr. Scholl's footpads, the complex history of a piece of leather, traces of bacterial action, and a few bits of the criminal himself. Then the dog meanders through the woods, snorting and evaluating. Suddenly a yelp. The pine needles, the dust, the people around him, the thousand smells of the forest floor—none of these interfere with his singular determination to follow the one faint odor imprinted on his brain. He will track that spoor wherever it leads— through creeks and swamps, across logs, down city sidewalks, up apartment stairs—one day, two days, even a week after the criminal has left the telltale bits.

The nose is also an organ of nostalgia. The smell of coffee, a whiff of briny seashore, the faintest trace of a certain perfume, or perhaps the etheric odor of a hospital room can stop you like a bullet. In a flash you relive a former moment, yanked backward in time by the fragrance stored inside your brain. I experience déjà vu whenever I visit India, a country that appreciates the sense of smell. In 1946, as a young doctor I sailed into the Bombay (now Mumbai) harbor after a twenty-three-year absence. An upsurge of distant childhood memories swept over me as the fantastic scents of that country drifted across the sea: steam-powered trains, bazaars, spicy food, sandalwood, Hindu incense, all airborne to my nose.

Nevertheless, a few days later these overpowering sensations faded into the background. The brain squelches odors after the initial excitement—"nasal ennui," Richard Selzer calls the phenomenon. Smell is primarily a sentinel warning and, once warned, why should the brain be troubled with redundancy? Fish merchants, tanners, garbage collectors, and sewage workers gratefully accept this mercy of habituation. "You get used to it," they say with total accuracy.

ORIGINS OF PLEASURE

Rods and cones, sound receptors, taste buds, olfactory cells—by serving the body as a whole, individual cells contribute to what I call the ecstasy of community. No scientist can yet measure how the sensation of pleasure materializes, but individual cells certainly play a role. Hormones and enzymes bathe the body's cells, bringing on the emotions' response of quickened breathing, a tremor of muscles, a flapping in the stomach.

If you search for a pleasure nerve in the human body, you will come away disappointed, for none exist. We have nerves to detect pain and cold and heat and touch, but no nerve dedicated to pleasure. Rather, the sensation emerges as a byproduct of cooperation by many cells.

I enjoy listening to a symphony orchestra. When I do, the chief source of what I interpret as pleasure centers in my ear, which picks up sound frequencies that flutter my eardrums as faintly as one billionth of a centimeter. My brain combines these impulses with other factors—how well I like classical music, my familiarity with the piece being played, the state of my digestion, the friends sitting beside me—and renders the result in a form I perceive as pleasure.

What about sexual pleasure? Even that is not as localized as we may think. Erogenous zones have no specialized pleasure nerves; the cells concentrated there also sense touch and pain. The saying "The sexiest organ resides between your ears" turns out to be true. Good sex draws on such things as romantic desire, a bank of intimate memories, visual delight, and perhaps the setting and background music. At a deeper, cellular level lies the urge to propagate life and ensure genetic survival. All these factors work together to produce the ecstasy of community.

Specialized cells have their origin in the fertilization of a single egg. In *The Medusa and the Snail,* author and physician Lewis Thomas muses about why people made such a fuss over the first "test-tube baby" in England. The true miracle, he says, is the union of a sperm and egg that results in another human being. "The mere existence of that cell," he writes,

> should be one of the greatest astonishments of the earth. People ought to be walking around all day, all through their waking hours, calling to each other in endless wonderment, talking of nothing except that cell. . . . If anyone does succeed in explaining it, within my lifetime, I will charter a skywriting airplane, maybe a whole fleet of them, and send them aloft to write one great exclamation point after another, around the whole sky, until all my money runs out.

From the basic protein of collagen, the maturing fetus fashions cells that divide up functions in exquisite ways: hair, skin, nails, bones, tendon, gut, cartilage, blood vessels. Billions of blood cells appear, millions of rods and cones—eventually some forty trillion cells emerge from a single fertilized ovum.

Alexander Tsiaras, a professor at the Yale Department of Medicine, filmed a video of the fetal stages from conception to birth, using MRI techniques that earned the inventor a Nobel Prize. The video compresses nine months of growth and development into a nine-minute film and is available on YouTube. As the video of sped-up fetal development plays, this mathematician drops his objectivity, awed by a system "so perfectly organized it's hard not to attribute divinity to it . . . the magic of the mechanisms inside each genetic structure saying exactly where that nerve cell should go."

One scene from the time-lapse video shows sixty thousand miles of capillaries and blood vessels taking shape where needed, following the genetic script built into a single cell. Aware of the intricate coding required to direct such a project, the programmer Tsiaras remarks,

> The complexity of the mathematical models of how these things are done is beyond human comprehension. Even though I am a mathematician, I look at this with marvel: How do these instruction sets not make mistakes as they build what is us? It's a mystery, it's magic, it's divinity.

A CHILD IS BORN

Finally, in the fullness of time a child is born. The umbilical cord stops pulsing and soon begins to shrivel. The drama of independent life is underway, and immediately the baby's cells join together in a cooperative response to the new environment. The baby's face recoils from the harsh lights and dry air; muscles limber up in jerky, awkward movements. Air rushes into lungs never before used, for oxygen filters through the lungs now, not the placenta. A team of bronchial passageways, diaphragm muscles, and all the other components of breathing must simultaneously lurch into motion.

The baby, though free and independent, is still incapable of supporting its own life. Happily, the mother's body has been readying itself for this new role since about age eleven. At puberty, a certain hormone present only in females begins to secrete at a gentle level. Today, in the body of every young woman millions of cells lie in wait, perusing the molecular structure of every hormone that happens by, much as one might scan one's email in search of an urgent call to action. All but a few of the body's cells ignore the chemical. Breast cells listen. They multiply and enlarge to shape the symmetry of a mature breast and then wait, quiescent, until pregnancy calls them to active duty.

The baby has no experience. It has never seen a breast and may, in fact, have never opened its eyes. Yet a baby instinctively knows what to do upon contact with a woman's breast. The baby creates suction by closing its mouth over an area of compliant skin and then contracting its throat muscles, while also shutting off the glottis to avoid drowning in the fluid. Nutritionists study with amazement the remarkable broth of vitamins, nutrients, antibodies, and macrophages that compose mother's milk. Oblivious, the baby knows only when and how to suck.

Soon the marvel of cooperation will unfold within the growing infant, whose hormones regulate development. Some body parts double in size, some triple, and some enlarge to hundreds of times their original size. What handicaps would result if the kneecap grew 10 percent faster than the tendons, ligaments, and muscles surrounding it, or if the right leg grew slightly longer than the left? Each body part grows in proportion to supporting structures, supplied with lengthened blood vessels at every stage of growth. The body's many parts work together in concert.

One can hardly avoid words like *miracle* and *marvel* when speaking of childbirth. Yet the phenomenon occurs so commonly that seven billion proofs now live on this planet. Within that clay-colored package of cells lies the origin of the ecstasy of community. The infant's life will include the joy of seeing his mother's delight at his first clumsy words, the discovery of his own unique talents and gifts, the fulfillment of sharing with other humans. Though a product of many cells, he is one organism. All his forty trillion cells know that.

ABBÉ PIERRE AND THE GOOD LIFE

I close my eyes and reflect on my life, sifting through memories to recall rare moments of intense pleasure and fulfillment. To my surprise, my mind passes by recollections of gourmet meals, vacations, and awards ceremonies. Instead, it settles on times when I have been able to work closely with a team in service to another human being.

On occasion that teamwork has helped to improve sight, arrest the crippling effects of leprosy, or save a leg from amputation. Sometimes those acts involved stress and apparent sacrifice. I have performed surgeries outdoors in primitive situations on a portable table in 110 degree heat with an assistant beside me holding a flashlight. Yet those times of working together, when we focused all our concentration on the goal of helping another, glow with an unusual luster. I was privileged to experience the ecstasy of community.

When Jesus described a fulfilling life, often his invitation sounded more like a warning than a sales pitch. "Count the cost," he said, and invited his followers to take on a yoke of service and to wash others' feet with a towel. While that attitude used to puzzle me, I now believe he was underscoring the need for individual cells to offer their resources in service to the whole Body. As Viktor Frankl wrote in *Man's Search for Meaning*, "Being human always points, and is directed, to something or someone, other than oneself—be it a meaning to fulfill or another human being to encounter. The more one forgets himself—by giving himself to a cause to serve or another person to love—the more human he is."

Although a spiritual Body following the Head may involve sacrifice, I have learned that service also opens up levels of personal fulfillment far exceeding any others. We are called to self-denial, not for its own sake but for a compensation we can obtain in no other way. Contemporary culture exalts self-fulfillment, self-discovery, and autonomy. In contrast, Jesus taught that only in losing my life will I truly find it. Only by committing myself as a "living sacrifice" to the larger Body will I find my true reason for being.

We sometimes think of sacrificial service with a self-focused sense of martyrdom. In fact, denying ourselves leads to a more abundant life. In the exchange, the advantage clearly rests on our side: crusty selfishness peels away to reveal the love of God expressed through our own hands, which in turn reshapes us into God's own image.

The value of service is better shown than told, and a powerful memory edges into my mind of an odd-looking Frenchman named Abbé Pierre. Unannounced, he showed up one day at the leprosy hospital at Vellore. A homely man with a big nose and a scraggly beard, he wore a simple monk's habit and carried a single carpetbag containing everything he possessed. I invited him to stay at my home, and there he told me his story.

Born into a noble family, as a teenager Pierre renounced his inheritance and gave away his possessions to charity. After becoming ordained as a Catholic priest, he served in the French Resistance, helping to rescue Jews from the Nazis. He spent a few terms in France's parliament until he became disillusioned with the slow pace of political change. With Paris still reeling from the effects of war and Nazi occupation, thousands of homeless beggars lived in the streets.

During one unusually harsh winter, many homeless Parisians froze to death. Pierre could not tolerate the endless debates by noblemen and politicians while so many street people starved outside. Failing to interest politicians in their plight, Abbé Pierre concluded he had only one recourse: to mobilize the beggars themselves.

First, he taught them to do their tasks more efficiently. Instead of sporadically collecting bottles and rags, they organized into teams to scour the city. Next, he led them to build a warehouse from discarded bricks and to start a business in which they sorted vast amounts of used bottles from big hotels and businesses. Then Pierre inspired each beggar by giving him the responsibility to help another beggar poorer than himself. The project caught fire, and within a few years an organization called Emmaus was founded to expand Pierre's work into other countries. The movement became known as "Abbé Pierre and the Ragpickers of Emmaus."

Now, he told me, after years of this work in Paris, there were no beggars left in that city. Pierre believed his organization was facing a serious crisis. "I must find somebody for my beggars to help!" he declared. That quest had brought him to Vellore.

He concluded by describing his dilemma. "If I don't find people worse off than my beggars, this movement could turn inward. They'll become a powerful, rich organization and the whole spiritual impact will be lost. They'll have no one to serve." As we walked out of the house toward the student hostel to have lunch, my head was ringing with Abbé Pierre's earnest plea for "somebody for my beggars to help!"

We had a tradition among the medical students at Vellore about which I forewarned all guests. Lunchtime guests would stand and say a few words about who they were and why they had come. Like students everywhere, ours were lighthearted and ornery, and they had developed an unspoken three-minute tolerance rule. If any guest talked longer than three minutes, the students would stamp their feet and silence the speaker.

On the day of Pierre's visit, he stood and I introduced him to the group. I could see the Indian students eyeing him quizzically—this small man wearing a peculiar old habit. Pierre started speaking in French, and a colleague named Heinz and I strained to translate what he was saying. Neither of us was well-practiced in French, and we could only break in now and then with a summary sentence.

Abbé Pierre began slowly but soon sped up, like an audio file playing too fast. I was on edge because I knew the students would soon shout down this great, humble man. Worse, I was failing miserably to translate his rapid-fire sentences. He had just visited the UN head-quarters where he had listened to dignitaries use fine-sounding, flowery words to insult other countries. Pierre was saying that you don't need language to express love, only to express hate. The language of love is what you *do*. He spoke even faster, gesticulating all the while, and Heinz and I looked at each other and shrugged helplessly.

Three minutes passed, and I stepped back and looked around the room. No one moved. The students gazed at Pierre with piercing black eyes, their faces rapt. He went on and on, and no one interrupted. After twenty minutes Pierre sat down, and immediately the students burst into the most tremendous ovation I had ever heard in that hall.

Completely mystified, I questioned the students afterward. "How did you understand? No one here speaks French."

One student answered me, "We did not need a language. We felt the presence of God and the presence of love."

Abbé Pierre had learned the discipline of loyal service that determines the Body's health. He had come to India in search of people more needy than his former beggars. He found them, some five thousand miles from his home, among our leprosy patients, many of whom were of the Untouchable caste and worse off in every way than his followers in France. Some visitors shied away from our patients; Abbé Pierre embraced them.

When he returned to Paris, the members of Emmaus worked with new energy, donating the proceeds to fund a ward at the hospital in Vellore. "No, no, it is you who have saved us," Pierre told the grateful recipients of his gift in India. "We must serve or we die."

In a fundamental human paradox, the more we reach out beyond ourselves, the more we are enriched and the more we grow in likeness to God—the Father of all good gifts. On the other hand, the more a person "incurves," to use Luther's word, the less human he or she becomes. Our need to give of ourselves in service to the whole Body is as great as anyone's need to receive.

PART THREE

OUTSIDE *and* INSIDE

Skin: *What is it, then, this seamless body-stocking, some two yards square, this our casing, our façade, that flushes, pales, perspires, glistens, glows, furrows, tingles, crawls, itches, pleasures, and pains us all our days, at once keeper of the organs within, and sensitive probe, adventurer into the world outside?*

Bone: *Bone is power. It is bone to which the soft parts cling, from which they are, helpless, strung and held aloft to the sun, lest man be but another slithering earth-noser.*

RICHARD SELZER

CHAPTER SEVEN

SKIN

The Organ of Sensitivity

As an intern in London I had the great privilege of training under Dr. Gwynne Williams, a surgeon who emphasized the human side of medicine. He strolled through the halls of our poorly heated hospital with his right arm Napoleonically tucked inside his coat, which, unknown to his patients, concealed a hot water bottle.

"You can't rely on what patients tell you about their intestines," Dr. Williams would admonish us interns. "Let their intestines talk to you." The hot water bottle made his hand a better listener. He taught us to kneel by a patient's bedside and gently slip a warm hand under the sheets onto the person's belly. "If you stand," he explained, "you'll tend to feel only with the downward-pointing fingertips. If you kneel, your full hand rests flat against the abdomen. Don't start moving it immediately. Just let it rest there."

We learned to anticipate a sudden tightening of the patient's abdominal muscles, a protective reflex. A cold hand guaranteed that those muscles would remain taut, whereas a warm, comforting hand coaxed them to relax. We gently caressed the abdomen, earning tactile trust. Once the muscles had slackened, we could sense the organs' movement in response to the simple act of breathing.

Dr. Williams was right. A trained hand exploring the abdomen can detect inflammation and the shape of tumors that more complicated procedures merely confirm. Touch is my most precious diagnostic tool.

Later, in India, I was asked on several occasions to examine female patients in Hindu or Muslim households that observed strict purdah. A woman would put her arm though a curtain and allow me to take her pulse; otherwise I could not see or touch any part of her body. From my four fingers resting on her wrist alone, I was expected to make a diagnosis. I felt handicapped, unable to listen directly to internal organs through my fingertips.

EVER-ADAPTING SKIN

Every small patch of skin has a different degree of sensitivity, and scientists have mapped the nerves as meticulously as Google has mapped the world. The physiologist Maximilian von Frey measured the threshold of touch, the amount of weight it takes for a person to sense that an object has come in contact with the skin. The soles of the feet, thickened for a daily regimen of abuse, do not report in until a weight of 250 milligrams is applied. The back of the forearm is triggered by 33 mg of pressure, the back of the hand by 12 mg. The really sensitive areas are the fingertips (3 mg) and the tip of the tongue (2 mg).

A wise mosquito will land on the forearm, not the sensitive hand, to go undetected. And only a foolhardy insect would attempt a stealth landing on soft lips.

The degree of sensitivity fits the function of that body part. Our fingertips, tongues, and lips are the portions of the body used in activities that require the most sensitivity. However, all touch sensors seem sluggish compared to those in the cornea of the eye, transparent, deprived of blood and thus extremely vulnerable. The cornea fires off a response if just two-tenths of a milligram of pressure is applied.

Moreover, the perception of touch changes constantly, based on context. A researcher lowers a 100 mg weight onto my forearm. Blindfolded, I realize that something is touching me. The sensation remains for four seconds, then fades. My brain now ignores the messages coming from nerve endings on my forearm, having decided there is no evident danger and no need to clog the circuits with useless information. I lose any awareness of the weight—that is, until the weight is removed, at which time my brain will draw attention to the change. Apart from this volume switch through which sensations pass, I could not wear wool or other coarse clothing; my body would incessantly remind me of its scratchy presence, and I could hardly concentrate on anything else.

I experience skin's adaptation whenever I lower myself into a hot bathtub. I run the water so hot that I can barely stand it and gradually lower my body, feeling at first as though I am easing myself into a bed of stinging nettles. Within ten seconds my skin adjusts, and the same water feels soothing and comfortable. I can continue raising the temperature of the water, and my body will adapt—up to a maximum point of 115°F, beyond which I will feel constant, nonadapting pain.

Bioengineers use the word *compliancy* to describe a material's capacity to mold to the shape of another surface, and skin exhibits this quality remarkably well. While trying to design shoes and tools for the insensitive feet and hands of leprosy patients, I have spent hundreds of hours studying the anatomy of living skin. Underneath the skin in the palm of the hand lie globules of fat with the look and consistency of tapioca pudding. So soft as to be almost fluid, fat globules cannot hold their own shape, and so they are surrounded by interwoven fibrils of collagen, like balloons caught in a rope net. The cheeks and the buttocks have more fat and less collagen, as anyone who has struggled with a double chin or sagging figure knows. In areas of stress, such as on the palm of the hand, fat is tightly sheathed by fibrous tissue in a design resembling fine Belgian lace.

I grasp a hammer in the palm of my hand. Each cluster of fat cells changes its shape in response to the pressure. It yields, yet cannot be pushed aside because of the firm collagen fibers around it. The resulting tissue, constantly shifting and quivering, becomes compliant, fitting its shape and its stress points to the precise shape of the handle of the hammer. Engineers nearly shout in awe when they analyze this amazing property, for they cannot design a material that so perfectly balances strength and pliability.

If my skin tissue were tougher, I might insensitively crush a goblet of fine crystal as I hold it in my hand; if softer, it would not allow a firm grip. When my hand surrounds an object—a ripe tomato, a hiking pole, a kitten, another hand—the fat and collagen redistribute themselves and assume a shape to comply with the object being grasped. This response spreads the area of contact, which prevents localized spots of high pressure.

In contrast, I have taken the hand of a human skeleton and wrapped it around a hammer. Against such a hard surface, the hammer handle will contact only about four pressure points. Without my compliant skin and its supporting tissues, those four pressure points would inflame and ulcerate after a few hammer blows. Because of compliancy, my entire skin-covered hand will absorb the impact.

Compliancy, a word with special meaning to my engineering colleagues, is a meaningful word for both the physical body and the spiritual Body. Compliant tissues covering my bones assume the shape—awkward or smooth—of whatever I am grasping. I do not demand that the object fit the shape of my hand; my hand adapts, distributing the pressure. The art of Christian living, I believe, can be glimpsed in this concept of compliancy. As my shape moves into contact with other, foreign shapes, how does my skin respond? Whose personality adapts? Do I, as does my grasping hand, become square to those things that are square, round to those things that are round?

It troubles me that Christians sometimes have a reputation for being divisive and exclusive. Though we live among others who may not share our beliefs and values, we have the clear example of Jesus, who found acceptance among physical and moral outcasts as well as despised minorities and Roman officers. Somehow he moved compliantly among diverse groups without compromising his good-news message of love and forgiveness.

The apostle Paul completes the analogy for us in 1 Corinthians 9, as paraphrased in *The Message*:

> Even though I am free of the demands and expectations of everyone, I
> have voluntarily become a servant to any and all in order to reach a wide
> range of people: religious, nonreligious, meticulous moralists, loose-living

immoralists, the defeated, the demoralized—whoever. I didn't take on their way of life. I kept my bearings in Christ—but I entered their world and tried to experience things from their point of view.

THE NEED FOR TOUCH

Dr. Harry F. Harlow loved to stand in his University of Wisconsin laboratory and watch the baby monkeys. He noticed that the monkeys showed a kind of emotional attachment to cloth pads lying in their cages. Intrigued, he watched as they caressed the cloths, cuddled next to them, and bonded to them much as a child bonds to a teddy bear. In fact, the monkeys raised in cages with cloth pads seemed healthier and less agitated than the monkeys raised in cages with wire-mesh floors. Was the softness, the *touchability* of the cloth making the difference?

Harlow constructed a surrogate mother out of terry cloth, with a light bulb behind it to radiate heat. His ingenious cloth mother featured a rubber nipple attached to a milk supply from which the babies could feed. They adopted her with great enthusiasm. Why not? She was always available and, unlike real mothers, never roughed them up or bit them or pushed them aside.

After demonstrating that babies could be "raised" by inanimate, surrogate mothers, Harlow next sought to measure the importance of the mother's touch. He put eight baby monkeys in a large cage that contained the terry-cloth mother plus a new mother, this one fashioned entirely from wire mesh. Harlow's assistants, controlling the milk flow, trained four of the babies to nurse from the terry-cloth mother and four from the wire-mesh mother. Each baby could get milk only from its designated mother.

A clear trend developed almost immediately. All eight babies spent their waking time huddled next to the terry-cloth mother. They hugged her, patted her, and perched on her. Monkeys assigned to the wire-mesh mother went to her only for feeding, then scooted back to the comfort and protection of the terry-cloth mother. When frightened, all eight would seek solace by clinging to the terry-cloth mother. A famous photo in introductory psychology books shows one of Harlow's baby monkeys clinging to the cloth mother with its hind legs while stretching mightily to feed from the tube on the wire mother.

Harlow concluded that young mammals need what he called *contact comfort* and will seek out whatever feels most like a real mother.

> We were not surprised to discover that contact comfort was an important basic affectional or love variable, but we did not expect it to overshadow so completely the variable of nursing; indeed the disparity is so great as to suggest that the primary function of nursing is that of insuring frequent and intimate body contact of the infant with the mother. Certainly, man cannot live by milk alone.

Anthropologist Ashley Montagu reported on these and many similar experiments in his elegant and seminal book *Touching*. He discovered that young animals require close physical contact with a mother for normal development. Except for humans, all mammals—think of dogs and cats—spend time licking their young. Animals will often die if they are not licked after birth; they never learn to eliminate waste, as one consequence. Montagu concluded that the licking provides essential tactile stimulation.

As pet owners know, animals do not outgrow the urge to be touched. A cat arches its back and brushes against its owner's leg. A dog wriggles on the carpet, begging for a belly rub. A monkey spends hours grooming and combing the hair on its fellow tribe members.

Montagu even suggested that human babies may need the tactile stimulations of labor. Only the human species goes through such a long, arduous birth process. Montagu believed the fourteen hours or so of uterine contractions, which have been described from the mother's viewpoint but never from the fetus's, may provide important stimuli to complete the maturation of certain bodily functions. Could this explain, he wondered, why babies delivered by Caesarean section have a higher mortality rate and a greater incidence of hyaline membrane disease?

Although the role of tactile stimulation during birth remains speculative, the need for touching after birth has been proved, decisively. As late as 1920, the death rate among infants in some foundling hospitals in the United States approached 100 percent, until Dr. Fritz Talbot of Boston introduced from Germany the unscientific-sounding concept of "tender loving care." While visiting the Children's Clinic in Düsseldorf, Talbot had noticed an old woman wandering through the hospital, always balancing a sickly baby on her hip. "That," said his guide, "is Old Anna. When we have done everything we can medically for a baby and it still is not doing well, we turn it over to Old Anna, and she cures it."

When Talbot proposed this novel idea to American institutions, administrators derided the notion that something as quaint as simple touching could improve their care. At that time, behaviorists were advising parents not to cuddle or coddle their babies; they proposed baby farms, where children could be raised by "scientific methods" away from their parents. The facts soon changed their minds. After Bellevue Hospital in New York adopted a policy that all babies must be picked up, carried around, and "mothered" several times a day, their infant mortality rate dropped from 35 percent to less than 10 percent. More recently, studies of children raised without touch in Romanian orphanages have shown a high incidence of stunted development, both physical and mental.

Despite these findings, even now touching is devalued and seldom viewed as essential for a baby's development. In general, the higher the social strata, the less frequently parents touch their infants. One study showed that fathers in the United States spend an average of thirty seconds per day in tactile contact with their children. Among some severely disturbed children, touching may represent the only hope for a cure. Autistic children in particular need persistent touching to coax them out of self-hugging isolation.

Montagu concluded that the skin ranks highest among the sense organs, higher even than eyes or ears. In addition to conveying information about the outside world, skin also perceives basic emotions. Am I loved and accepted? Is the world secure or hostile? The skin osmotically absorbs that primal assurance.

Touch words have edged into our vocabulary as expressions of how we relate to others. We rub people the wrong way, or conversely we give them strokes. One person represents a soft touch; another, we handle with kid gloves. We are thin-skinned, thick-skinned; we get under each other's skins. We relate tactfully or tactlessly.

Touching involves risk. It can evoke the cold, armor-like resistance of a hurt spouse refusing to be comforted or the lonely shrug of a child who insists, "Leave me alone!" Yet it can also conduct the electric tingling of love-making, the symbiosis of touching and being touched simultaneously. A kiss, a slap on the cheek—both are forms of touch, and both communicate. Skin cells offer a direct path into the deep reservoir of emotion we metaphorically call "the human heart."

LOVE'S MEDIUM

The skin of a spiritual Body too is an organ of communication: our medium of expressing love. I reflect back on how Jesus acted on earth. His hands reached out to touch the eyes of the blind, the skin of the person with leprosy, and the legs of the lame. When a woman pressed against him in a crowd to tap into his healing energy, he felt the drain of that energy, halting the crowd to ask, "Who touched me?" His touch transmitted power.

I have sometimes wondered why Jesus so frequently touched the people he healed, many of whom must have been unattractive, obviously diseased, unsanitary, smelly. He could have waved a magic wand, which would have affected more people than he could personally touch. He could have divided the crowd into affinity groups and organized his miracles—paralyzed people over there, feverish people here, people with leprosy there—raising his hands to heal each group efficiently, en masse. Instead, he chose a different style.

Jesus' mission was not chiefly a crusade against disease (if so, why did he leave so many unhealed in the world and tell followers to hush up details of his miracles?) but rather a ministry to individual people, some of whom happened to have a disease. He wanted those people, one by one, to feel his love and compassion. Jesus knew he could not readily demonstrate love to a crowd, for love usually involves touching.

I have mentioned the need for us as Jesus' followers to share resources such as food and medicine with those in need. Having participated in such activity overseas, I am convinced that we best express such love person to person, through touch. The further we remove ourselves from personal contact with the needy, the further we stray from the ministry Jesus modeled for us.

In India, when I would treat a serious case and prescribe a treatment, sometimes the relatives of the patient would go and purchase the medicine, then bring it back and ask me to give it to the patient "with my good hands." They believed medicine had more power to help the patient if it came from the hands of a physician. And in the United States, recent studies on touch show that a doctor's comforting touch can leave patients with the impression that a visit lasted much longer.

When I left India, I moved to the grounds of the only leprosarium in the continental United States. Carville has a poignant history. The hospital began in the 1890s when an order of Catholic nuns, the Daughters of Charity of St. Vincent DePaul, felt a specific calling to serve leprosy patients. Because no one wanted to live near a leprosarium, they purchased a remote plot of swampland on the Mississippi River under the guise of establishing an ostrich farm. The first patients, blackened and hiding under tarpaulins, were smuggled in at night on coal barges.

News of the leprosarium soon leaked, however, and the construction workers slipped away. Misconceptions of the disease struck such fear that no one would risk exposure to it. Undeterred, the nuns pursued their calling. Under the direction of a stout and courageous Mother Superior, they took up the hoes and shovels themselves, digging canals to drain the swamp. With no prior construction experience, teams of sisters in starched, sweltering habits dug foundations and erected buildings. Only they cared enough to touch and treat the disfigured patients who came to them in the dark of night.

Nearly a century later, I find myself treating leprosy patients at that same hospital. Many of them, their nerve cells destroyed, cannot distinguish what they touch: furniture, fabric, grass, asphalt—it all feels the same. When they put their hands on a hot stove because it feels no different from a cool one, I must treat their damaged hands.

I hate leprosy. Untreated, the disease slowly silences nerve sensors on the hands and feet until the afflicted lose the ability to sense human contact. Many cannot even sense when another person holds their hands or caresses them. Because of ignorance and superstition, this disease destroys social contact between patients and their friends, employers, and neighbors. Leprosy is a devastatingly lonely disease.

As at Carville, many of the great advances in leprosy research and treatment have come about because of Christian action, especially by The Leprosy Mission (British) and its counterpart, American Leprosy Missions. I have sometimes wondered why leprosy merits its own task force; I know of no Malaria Mission or Cholera Mission. Perhaps the reason traces back to the leprosy patients' starvation for human touch. Theirs is a unique and terrible privation, and Christian love and sensitivity meet it best.

Medical teams can offer great assistance to leprosy patients. Physicians treat the raw ulcers and painstakingly reconstruct feet and hands through tendon transfers and plastic surgery. We transplant new eyebrows to replace missing ones, repair useless eyelids, and sometimes even restore sight. By training patients in constructive jobs, occupational therapists give them new life. Yet of all the gifts we can give to leprosy patients, the one they value most is the gift of being touched. We don't shrink away. We love them skin to skin.

When I first began working with leprosy, I did so with trepidation. A missionary physiotherapist, Ruth Thomas, helped me overcome my latent fear. Ruth set up a physiotherapy area in our clinic, equipping it for hot paraffin treatment and electrical stimulation of muscles. She also believed that vigorous hand-to-hand massage would help prevent hands from stiffening. Every day she sat in the corner stroking the hands of leprosy patients. "Ruth, this is intimate skin-to-skin contact!" I warned her. "You really should be wearing gloves." She would smile, nod, and keep on stroking. Ruth Thomas achieved remarkable success with her simple therapy, success that I credit as much to her gift of human touch as to any massage technique.

A few months after we opened the unit, I was examining the hands of a bright young man, trying to explain to him in my broken Tamil that we could halt the progress of the disease, and perhaps restore some movement to his hand, though we could do little about his facial deformities. I joked a bit, laying my hand on his shoulder. "Your face is not so bad," I said with a wink, "and it shouldn't get any worse if you take the medication. After all, we men don't worry so much about faces. It's the women who fret over every bump and wrinkle." I expected him to smile in response, but instead he began to shake with muffled sobs.

"Have I said something wrong?" I asked my assistant in English. "Did he misunderstand me?" She quizzed him in a spurt of Tamil and reported, "No, doctor. He says he is crying because you put your hand around his shoulder. Until he came here no one had touched him for many years."

CHAPTER EIGHT

THE VISIBLE YOU

WHILE LEPROSY RESEARCH CONSUMED MY TIME in India, my wife, Margaret, trained in ophthalmology and became an expert eye surgeon. Because many of the neediest people could not travel to the hospital, her team organized a mobile unit that made monthly circuits into rural areas. On a given day a designated building, perhaps a school or an old rice mill, would receive a stream of villagers afflicted with eye problems. If there was no building available, the team would set up portable tables under a banyan tree. Sometimes two doctors performed a hundred cataract operations in a single day.

One year Margaret's team staffed a camp for several weeks in a region devastated by drought. Desperate villagers straggled in from all directions. Some of them begged for needless surgery—even asking doctors to remove one of their eyes—in exchange for food.

As her helper at that hectic camp, Margaret chose a shy boy about twelve years old. He stood on a box, with a baggy hospital gown wrapped around him, charged with strict orders to hold a three-battery flashlight so that the light beamed directly on the cornea of the patient's eye. Margaret wondered how an inexperienced village boy would cope, watching people's eyes sliced open and stitched together again.

The child performed his task with remarkable aplomb. During the first five operations he scrupulously followed Margaret's instructions on when to shift the angle of light, aiming the beam with a steady, confident hand. During the sixth case, however, he faltered. Margaret kept saying softly, "Little brother, shine the light properly," which he would momentarily do. Soon the beam would again bob away from where she was cutting. Sensing that the boy simply could not bear to watch, Margaret paused the procedure and asked if he was feeling well.

Tears ran down his cheeks and he stammered, "Oh, d-doctor, I-I cannot look. This one, she is my mother." Margaret found another temporary helper in order to give the boy a break. Later in the day, she learned that his mother had gone blind as an adult, shortly before her son's birth. Now she was hoping to regain some sight.

Ten days later the woman's stitches were removed, and the team gave her eyeglasses. She first tried to blink away the dazzling light, gradually learned to adjust, and then for the first time ever she focused on the face of her son. A smile creased her face as she reached out to touch him and pull him close. "My son," she said, "I thought I knew you, but today I see you."

THE BODY'S CANVAS

Sometimes I envy my wife's field of medicine, for eyes are our only direct portal to what goes on inside the body. Without invasive surgery, an ophthalmologist can observe firsthand healthy or diseased cells. The rest of us, like the mother of Margaret's young helper, see only the organ of skin. We perceive, and make judgments of, another person by the body's surface.

We humans have a love affair with skin, and our chief response is to adorn it. Men shave off a night's facial growth from hair follicles, rearrange other hairs atop their heads, perhaps worry over a pimple, and inspect a mole or two. Women expand the ritual, powdering dry the dense oil glands on the nose, curling or plucking hairs around the eye, and daubing the skin like an artist's canvas with makeup, eye shadow, and lipstick. Then, alone among all God's creatures on earth, we sense the need to swathe the skin, supporting a multibillion-dollar fashion industry in the process.

Although skin is opaque, not transparent like the eye, those of us in medicine learn to read it as a projection of the health within. Anemia casts a ghostly pallor over its victims; jaundice yellows the skin; a form of diabetes shades it bronze. An allergist can crack the secret code of your body's likes and dislikes by mapping a grid on your back and pricking the skin with tiny potions. Does the problem stem from dog hair? Pollen? Shellfish? Your skin will unriddle the mysterious vomiting or sneezing.

Skin also gives clues to the emotional world within. Possessing few voluntary muscles on our skin, we humans cannot twitch it at will, as can a horse. Yet we do have mastery over the face, which expresses true feelings. A slight puffiness or downward curve of the lips can warn a spouse to walk on eggshells. Like scars on a tree trunk, facial contours may betray emotional wounds from childhood.

Skin is endlessly adaptable, smooth as a baby's stomach here and rough like a crocodile there. It flexes and folds around joints, facial bones, gnarled toes, and fleshy buttocks. View sections of scalp, lip, nipple, heel, abdomen, and fingertip through a laboratory microscope, and you can hardly believe they come from the same organ. Form and function go together: the tiny ridges crisscrossing the surface of our fingers increase the ability to grip, much like the tread on a snow tire. Yet each of us has a different pattern for these ridges, an embellishment that the FBI exploits with its fingerprint files.

In comparison with other species, our outward appearance seems dull. We have nothing to compete with the design features of a scarlet macaw or killer whale or giraffe, not to mention the tropical fish found on coral reefs. Human skin is furless and nearly monochrome: shaded to yellow, brown, black, white, and red, to be sure, but unvarying across the body save for a darker palette on lips and nipples. Even so, artists since the days of cave dwellers have found an unending subject of fascination in plain human skin.

A GREAT WALL OF DEFENSE

We give inordinate attention to skin's appearance, hardly its most crucial contribution. Skin forms a protective barrier, a Great Wall that keeps the inside in and the outside out. Sixty percent of the body consists of fluids, and these would soon evaporate if deprived of skin's covering. Absent skin, a bath or dip in the pool would prove fatal, for foreign fluids would rush in to dilute the blood and waterlog the lungs.

Civilization taxes skin's capacities. On a given weekend we might dig in the garden gloveless, spill kerosene on our hands at the barbecue grill, and clean paint brushes with turpentine, then scour it all off with abrasive powder and a roughened pad. Somehow skin survives.

If I clean out a muddy gutter with my bare hand or reach in to unstop a toilet, I'll encounter a swarm of bacteria, yet my skin cells loyally guard against their entering my body. Each of us carries as many bacteria on our skin as there are people inhabiting this planet, and skin uses chemicals, electro-negative charges, and an army of defending cells to keep the marauders at bay.

Larger animals make their home in skin's fissures. For most of human history mites, fleas, bedbugs, and lice were an accepted part of skin's landscape. Thomas à Becket's hair shirt was studded with lice, and Samuel Pepys had to return a wig that came from the hairdresser full of nits. Even today an eight-legged creature just a third of a millimeter long, the *Demodex folliculorum,* burrows its way alongside hair shafts and contentedly lives out its days in its tunnel of choice, the eyelash follicle. Ophthalmologists find this cigar-shaped mite, apparently harmless, on almost every person they examine. Male and female Demodexes merrily mate in their tunnel, and as many as twenty-five of the creatures can congregate in one warm, oily fat gland.

It's a rough world outside, and as casualties mount among skin cells, the epidermis provides a continuous stream of reinforcements. Dying cells curl like cornflakes, ready to slough off and make room for fresh replacements underneath. Anatomists who count such things estimate we lose several million skin cells a day. Merely shaking hands or turning a doorknob can produce a shower of a thousand skin cells; who can calculate the loss during a game of tennis?

Though we leave a trail of shed skin cells wherever we go, many of them stay close to home. Up to 90 percent of all household dust consists of dead skin—friendly scrapings of you, your family, and your guests, waiting to be removed with a dust cloth or vacuum cleaner without a moment of gratitude for their sacrifice.

HYPERSENSITIVITY

As our first line of defense, skin must respond to stresses that are constantly changing. While researching how best to protect the skin of leprosy patients, I gained great admiration for skin's adaptive properties.

We developed a mechanical device that presses a metal rod against the fingertips with measured force. If I put my hand under the tiny hammer, it feels rather pleasant, like a vibromassage. But if I let the machine run on for several hundred beats, my finger turns slightly red and feels uncomfortable. After fifteen hundred beats I must pull my finger out, for I can no longer bear the pain. When I return to the machine the next day, I can only tolerate a couple of hundred beats before yanking my finger away.

Mild inflammation brings on a condition called hypersensitivity. My finger feels warm as blood surges to the point of stress, and swollen as the body cushions it with extra fluid. The same finger that endured many blows from a tiny hammer yesterday has become hypersensitive, and in its inflamed state just a few more thumps could lead to a blister or ulcer.

Likewise, a burned finger becomes hypersensitive to heat. More than once I have put my hands in a basin only to discover that my hands are sending mixed signals. My left hand reports that the water is hot while my right hand says warm. Then I remember an incident from breakfast: a drop of hot bacon grease popped out of the pan and landed on my left hand. Nerve endings at that spot lowered their tolerance threshold and are now reporting warm water as hot because even a little heat might harm the mildly inflamed tissues.

Who has not felt the irritation of a sore finger always seeming to get bumped every few minutes no matter how careful you are. That phenomenon has a sound physiological basis: pain cells near the site of injury have become ten times more sensitive to pain. In effect, pain cells "turn up the volume" so that I won't foolishly subject my skin to more hot grease or hammer blows. In these remarkable ways, hypersensitivity produces a shield of protection around vulnerable areas.

At times all of us experience a psychological form of hypersensitivity. An accumulation of small stresses builds up—overdue bills, work pressures, house repairs, irritating habits of family members—and suddenly every minor frustration hits like a sledgehammer. Pain, whether physical or emotional, works precisely because it is loud and insistent. Hypersensitivity alerts the body to skin's urgent need of relief from stress. Likewise, emotional hypersensitivity in one member can alert the larger community to the need for respite or outside care.

A healthy Body feels the pain of its weakest parts. We may be called on to bolster a bruised ego, mediate a dispute, or take upon ourselves some of the minor stresses that have piled on others. As a former missionary, I cannot overstate the life-sustaining role of people back home who supported me by praying and writing letters. These unusually sensitive cells in the Body sought out my hardships and nourished me in times of need. Such dedicated people make the difference between a missionary or aid worker who serves twenty years and one who breaks down after a short time.

Once I served as a medical officer at a professional boxing match, assigned to treat the injuries that occurred during the match. I only accepted the assignment once, for the sight of two men pounding living cells to destruction offended all my medical sensibilities. One memory especially stays with me. The trainer of one of the heavyweight boxers rushed to the boxer slumped on a stool near where I stood. "The left eyebrow!" he yelled excitedly, pointing to his own dilated eye for emphasis. "Pound him on the left eye! You've landed some good ones—it's already swelling. A couple more jabs and you'll bust it open!"

The boxer followed his instructions, targeting the inflamed, hypersensitive lump above his opponent's eye. After the fight I had to sew up the remains of skin and eyebrow. The relentless pummeling had taken its toll.

That scene has come to my mind in very different circumstances—such as a dinner party at a friend's home. Everyone is conversing amiably until the husband says something to his wife that seems slightly charged. The wife flushes with obvious embarrassment, and the husband appears a bit smug. A blow, genteel but deadly, has landed. Dinner proceeds with some awkwardness after the comment. In such an interchange—a remark, slightly veiled in humor, about housecleaning, some past disagreement, a personal habit, sexual performance, or in-laws—I hear replayed, "Hit him again. The left eyebrow!"

I have seen the same pattern of destruction in other groups: church members gossip about their pastor, an employer mercilessly harasses a conscientious employee, parents or siblings needle a clumsy child. Where is the gracious adaptation that accommodates to weakness? All of us could take a lesson from the human body's adaptations to stress. "Carry each other's burdens," said Paul, "and in this way you will fulfill the law of Christ" (Galatians 6:2).

DISTRIBUTING STRESS

The human body gives a striking model of how to respond to hypersensitivity: by redistributing stress.

I go through this process whenever I buy new shoes. Although the loafers felt comfortable during my test paces in the store, when I walk a mile home a friction spot on my foot begins to call for help. If I continue walking, my body will redistribute stress away from the tender spot: I limp. The new gait, though awkward and unnatural, gives relief to the vulnerable tissues.

Without stress distribution, our daily activities would be fraught with danger. I know, for I have treated scores of leprosy patients who will never walk again because a defective pain system failed to warn the brain to redistribute stress. Feeling no pain, they tend to walk on the very same part of the foot, rather than spreading out the stress across the foot's surface.

Even rest can represent danger. Nurses and caregivers must stay vigilant for bedsores, which develop when a person lies on the same area of skin without moving. If the body hears a whimper of pain, it will turn a bit, distributing the stress to other cells. But in paraplegics or those made insensitive by leprosy or diabetes, horrible bedsores may result. I thank God for the millions of sensors embedded in my skin that tell me when to shift weight in my buttocks or reposition my legs or back, or change my gait when walking.

When I turn from the physical to the spiritual Body, I see the need for a similar adaptation. A spiritual Body must constantly evaluate which parts need special attention. Outer, frontline cells will require qualities of resiliency and firmness, while inner cells need refuge in order to lead lives of quiet contemplation.

From my own observation, the church tends to fail at this principle of redistributing stress in two crucial areas. First, when we put leaders on the frontlines—pastors, priests, missionaries, other public representatives—we demand too much. We have unrealistic expectations, giving them no chance to "limp." I caution such leaders to surround themselves with sensitive friends and associates who can detect danger signs and pursue ways to redistribute those pressures.

My laboratory research proved that the subtler forces of *repetitive* stress hold greater peril for my patients than the obvious hazards of laceration or burning. We must not overlook the cumulative effect of everyday stress in the lives of pastors: incessant phone calls, a nettlesome board, financial pressures, the burdens of counseling, loneliness, social ostracism. These represent far greater dangers than more visible crises in the church.

The church could learn a second major lesson from the human body: certain members need protection at vulnerable times, especially during their spiritual infancy. In the United States, especially, I have seen a pattern of showcasing new converts such as athletes, politicians, musicians, actors, and beauty queens. These enthusiastic new recruits may capture the attention of the media for a short time. After trying to project the image expected of them, some of them later abandon their faith in bitterness and disgust.

The syndrome has a parallel in the human body: the skin disease known as psoriasis. The disease has one cause: skin cells that normally take three weeks to migrate to the surface force their way up in a few days. Those immature cells arrive unprepared for the stresses of light, ultraviolet rays, temperature, and atmosphere on the surface. They die quick, ugly deaths, scarring their miserable victims. Is there not a lesson here for the Christian subculture that insists on forcing newly converted celebrities into the spotlight without allowing time for spiritual maturity to develop?

When it works well, the Body of Christ can surround a vulnerable person and wisely redistribute stress. I think of a divorced woman I know in a small church. After her husband left her for another woman, she struggled to hold her life together. Burdened by feelings of guilt and rejection over his leaving, she also had to cope with four children, an empty bank account, and a house in poor repair. Her local church community responded in loving and practical ways: by babysitting, painting the house, repairing the car, inviting her to special events. Today, five years later, she still leans on the church to help her cope. She has regained much health because church members, like cells in a body, surrounded her with their strength and relieved the pressures that could have destroyed her.

BLISTER AND CALLUS

On a break from medical school, I spent one summer sailing on an eighty-foot schooner on the North Sea. The first week, as I pulled on heavy ropes to hoist the sail, my fingertips became so sore that they bled and kept me awake at night with the pain. By the end of the second week calluses were forming, and eventually thick calluses covered my fingers. I had no more trouble with tenderness that summer, for the calluses protected me. But when I returned to medical school, I found to my chagrin that I had lost the finer skills required in dissection. The calluses made my fingers so insensitive that I could scarcely feel the instruments. I panicked, worried that those thick pads of callus had ruined my career as a surgeon. In time, however, when my body sensed I had no need for the extra protection, it shed the layers as gladly as a molting insect sheds its shell. Sensitivity returned.

The pattern repeats itself today when I ignore warning signs while using the spade in my garden. If I work too long, the top layer of epidermis separates from the layers underneath it and mushrooms up to form a perfect dome supported by the sudden influx of liquid: a blister. A temporary measure, the blister cools the area, cushions shock, and disperses stress—in short, it gets me through the day. Human beings have bad habits, though: we tend to repeat, over and over, the very stresses that bring on inflammation, hypersensitivity, and blisters. A tennis player will work through five consecutive blisters before she convinces her body of the need to develop a more permanent adaptation: skin then alters a blister into a callus.

Skin struggles to find the proper balance between hypersensitivity and callus. Similarly, those on the frontlines of ministry, the "skin" on the Body, expose themselves to ever-changing stresses. Sometimes a person in ministry needs the fine skill of a surgeon, for the repair of human souls may require more sensitivity than the repair of human bodies. At other times the person in ministry, overburdened, short of resources, besieged by unsolvable problems, needs a layer of callus.

I compare my field trips out into Indian villages, where hundreds of patients lined up for treatment, to the situation at Carville, where we have as many staff members as patients. The slower pace at Carville allows me time to research problems in-depth and to get to know each individual patient. On the Indian field trips, I had to forfeit that personal sensitivity for the greater demands of efficient medical procedures. I could not possibly have gotten personally involved with each of the patients.

Medical staff, social workers, and counselors who work amid overwhelming human needs must sometimes develop a protective callus. Young doctors ask my advice on how to cope with these pressures without giving in to hardness and cynicism. I tell them to pray daily, asking God to identify one or two select patients with special needs. I cannot be equally sensitive to everyone, and I must not grow insensitive to all. Rather, I depend on the Spirit to help me sense those who need something beyond strictly medical care.

Wheaton College in Illinois has formed a Humanitarian Disaster Institute to study the effect of disasters on aid workers as well as victims. What emotional and spiritual care helps cultivate a spirit of resilience? How can aid workers keep from being overwhelmed by trauma? Christian ministry dangles on a pendulum between hypersensitivity and callus. Some workers remain so hypersensitive to the pain around them that they succumb to burnout, having failed to develop layers of protective callus. Others develop a callousness that keeps them from having true empathy for the victims of disaster.

Those of us in support roles in the Body must accept the responsibility of carefully monitoring our representatives on the frontlines. We cannot expose them to unrelieved human sorrow, and these servants must rely on others to help convince them to pull back or to shift their load to someone else. Too much sensitivity—or too little—can immobilize either a physical body or a corporate one.

SERVING ON THE FRONTLINES

In England I once spoke to a congregation in a centuries-old stone church. I noticed that the stained-glass windows portrayed saints with their hands together before them, almost in an attitude of prayer. It occurred to me that in normal life I rarely see people with their hands together like that—except in the operating room. After I scrub, if the patient is not ready, I join my hands fingertip to fingertip lest I touch anything in the room that might harbor germs. What a terrible image for saints! We should have our hands outstretched, embracing a contaminated world because we trust the protection of loyal skin cells in the spiritual Body.

Some brave souls do find themselves on the frontlines: members of the persecuted church, relief workers in war zones and refugee camps and disaster areas, health workers fighting disease in difficult places. From the rest of us, they deserve support and prayer, for life on the surface of the Body is never easy.

I think of my own mother, from a society home in suburban London, who went to India as a missionary. When Granny Brand, as she was universally called, reached sixty-nine, the mission asked her to retire. She did . . . until she found a new range of mountains where no missionary had ever visited. Without her mission's support, she climbed those mountains, built a little wooden shack, and worked another twenty-six years. Because of a broken hip and creeping paralysis, she could only walk with the aid of two bamboo sticks, but on the back of an old horse she rode all over the mountains, a medicine box strapped behind her. She sought out the unwanted and the unlovely—the sick, the maimed, and the blind—and brought them treatment. When she came to settlements who knew her, a great crowd of people would rush out to greet her.

My mother died in 1974 at the age of ninety-five. Poor nutrition and failing health had swollen her joints and made her gaunt and fragile. She had stopped caring about her personal appearance long ago. The villagers in that mountain range, though, saw beauty in her leathery, wrinkled skin. She was part of the advance guard, the frontline presenting God's love to deprived people.

In the spiritual Body, skin represents the membrane lining that defines our community and expresses God's presence in the world. The watching world sees our skin, and Jesus stated clearly what we should make visible to a watching world: "By this everyone will know that you are my disciples, if you *love* one another" (John 13:35, emphasis added). Skin—soft, warm, touchable—conveys the essence of a God who is eager to relate to us in love.

As the world makes initial contact with the church, what is its texture, its appearance and "feel"—its skin? Do people see "love, joy, peace, forbearance, kindness, goodness, faithfulness, gentleness and self-control" (Galatians 5:22-23)? We judge people by appearance, studying facial expressions for some hint of mood or glimpse into them. In the same way, we as a Body are being scrutinized as others draw conclusions about God from our appearance.

CHAPTER NINE

The MOST TRUSTWORTHY SENSE

JUST AS I REACHED THE PODIUM to deliver an address, I began to feel feverish and nauseated.

I was in New York on the final leg of a tour across the United States funded by the Rockefeller Foundation. I had visited renowned hand surgeons and pathologists to investigate why leprosy causes paralysis. Now I faced one last assignment, a scheduled lecture before the American Leprosy Mission.

Although I managed to get through the talk, the fever continued to rise as I made my way to the subway station. At one point during the journey I swayed and fell to the floor of the subway car, too dizzy to sit or stand. Other passengers ignored me, no doubt assuming I was inebriated.

Somehow I staggered to my hotel. In a fog, I dully realized I should call a doctor, but the hotel room had no telephone. The illness so overwhelmed me that I could only curl up on the bed and moan. For several days I remained there, aided by a kind bellboy who daily fetched me orange juice, milk, and aspirin.

Though weak and unsteady, I recovered enough to make my ship's voyage back to England. I spent most of the time in my cabin, resting to regain strength. After we docked at Southampton, I rode a train to London, sitting in a cramped corner, hunched over, and wishing the interminable trip would end.

At last I arrived at my aunt's house, emotionally and physically drained. I collapsed like a sack of potatoes into a bedside chair and removed my shoes. Then came probably the darkest moment of my entire life. As I leaned forward and pulled off my socks, I became aware that my left heel had no feeling. A dread fear worse than nausea seized my stomach. After working with leprosy patients, had it finally happened? Was I now to become one of them?

I stood stiffly, found a straight pin, and sat down again. I lightly pricked a small spot of skin below my ankle and felt no pain. I jabbed the pin deeper, longing for a reflex, but there was none—just a speck of blood oozing out the pinhole. I put my face between my hands and shuddered, longing for pain that would not come.

For seven years my team and I had battled against centuries of tradition to gain new dignity and freedom for leprosy patients. We had proudly torn down the barbed-wire fence surrounding the leprosy village at Vellore. Now, images of my patients' faces ravaged by the disease filled my mind. Was this to be my future?

I had assured staff members that leprosy was the least infectious of all communicable diseases and that proper hygiene would practically guarantee they would not contract the disease. Now I, their leader—a *leper.* That vicious word I had banned from my vocabulary rose up accusingly. How glibly had I urged patients to overcome stigma and prejudice.

My thoughts and emotions churned. I would need to separate myself from my family, of course—the children of patients were the most susceptible to infection. Perhaps I should stay in England rather than return to India. But what if word leaked out? I could envision the headlines. And what would happen to our leprosy work? How many health workers would continue in view of the risk?

I lay on my bed all night, fully clothed except for shoes and socks, sweating and breathing heavily from tension. I pictured the disease spreading across my face, my feet, my fingers. Scenes flickered through my mind, vivid reminders of what I would lose as a leprosy patient.

Hands were my stock in trade, and my career as a surgeon would soon end. How could I use a scalpel, lacking fine finger control and response to pressure? Much else would slip away. Feathers, a dog's fur, silk, wool— touch filled my world, and because I treated leprosy patients who had lost most of these sensations, I cherished them.

Dawn finally came and I arose, unrested and full of despair. For a moment I stared in a mirror, summoning up courage, then picked up the pin again to map out the affected area. I took a deep breath, jabbed in the point—and yelped aloud. Never have I felt a sensation as welcome as that live, electric jolt of pain synapsing through my body. I fell on my knees in gratitude to God.

Soon I was laughing sheepishly and shaking my head at my foolishness of the night before. In retrospect, everything made perfect sense. As I had sat on the train, weakened enough to forgo the usual restless motion of muscles in a cramped place, I had numbed a nerve in my leg. And then, exhausted, I had exaggerated my fears and jumped to a false conclusion. There was no leprosy, only a tired, anxious traveler recovering from influenza.

A WORLD OF TOUCH

That dismal affair, which I was too ashamed to mention to anyone for years, taught me a lasting lesson about perception. Since then I have resolved to feel, *really* feel the sensations of touch. Forests, animals, fabric, sculpture—these beg for exploration by sense-hungry fingertips. Sensors on the body's surface hum with reports on the surrounding environment.

Rolled thin like pie dough, the skin abounds with half a million tiny transmitters reporting their discoveries. Think of all the stimuli your skin monitors each day: wind, particles, parasites, pressure, temperature, humidity, light, radiation. Skin has the toughness to withstand the rigorous pounding of a marathon run, yet the sensitivity to detect a mosquito landing on it. One tap of the fingernail can tell me if I am touching paper, steel, wood, plastic, or fabric.

Of all the senses, touch is the most trustworthy. Give a baby an object to play with and she will finger it, then bring it to her mouth and tongue it. To her, touch is primary, and auditory and visual senses are secondary. Later, she may touch a magician's props to see if they are real; she cannot trust her eyes. A mirage may fool the eye and the brain but not the skin's touch. And even adults trust tactile senses more readily, hence "tangible" proofs. The disciple Thomas doubted visual reports of Jesus' resurrection, declaring, "Unless I see the nail marks in his hands and put my finger where the nails were, and put my hand into his side, I will not believe" (John 20:25).

I recall when my daughter Mary, three years old, was trying to overcome a fear of thunderstorms. Although she believed we were safe inside our house, as the lightning flashed closer and closer, she ran to me and put her small hand in mine. "We aren't afraid, are we, Daddy?" she said in a wavering, uncertain voice. Just then a tremendous clap of thunder crashed nearby and all the lights went out. Mary, breathing in short gasps, cried out more urgently, "Daddy! We aren't afraid, are we?" In her little hand, trembling with fear, I could read past her brave words to her true state. Skin communicates to skin.

In many dictionaries the definition for *touch* runs the longest of any entry. I can hardly think of a human activity—sports, music, art, cooking, mechanics, sex—that does not rely on touch. It is the most alert of our senses when we sleep and the one that seems to invigorate us when awake: the lovers' embrace, cuddling a baby, a contented sigh after a massage, the sting of a hot shower. Helen Keller, blind and deaf and yet a *cum laude* graduate of Radcliffe and author of twelve books, shows what the brain can accomplish with input from touch alone.

From the skin, I better understand one requirement on the frontlines of a spiritual Body: to sensitively perceive the people we contact. Beginning counselors, eager to help people, must remind themselves, "First, you must listen. Your wise advice will do no good unless you begin by listening carefully to the person in need." Skin provides a more basic kind of perception than what passes through the eye or ear—a tactile perception compiled from thousands of sensors.

If skin sensors detect a minute change in air pressure or temperature, they fire off messages to the brain. In the same way, followers of Jesus, "in the world but not of the world," encounter a constant stream of signals about the environment around them. The Body is universal, and its sensors report in from lakeshore apartments in Chicago, the slums of Nairobi, the jungles of Peru and Sri Lanka, the deserts of Russia and Arabia. In earlier days, foreign missionaries did not always sense the worth and beauty already present in different cultures. Today, Christian missions are more sensitive to culture, and to physical and social as well as spiritual needs. The best, most effective kind of love begins with a quiet attention, a tactile awareness that senses a need and responds appropriately and personally.

I do not believe humanitarian or mission work necessarily becomes more effective as it grows more specialized. Advances in technology may offer benefits, for example in medical work, but I have seen Christian medical agencies in India gradually lose their original mission as they become institutions with buildings and staff to support. The quality of treatment rises, but so does the expense, and to sustain the work they then focus on techniques that attract patients who can pay. Meanwhile the poor and unloved, unable to afford the mission hospital, must turn to a government clinic for help.

TACTILE AWARENESS

In contrast, I recall the impact my parents had in a remote, mountainous region of India. Although they went to India to preach the gospel, by living in tactile awareness of people's needs they began to respond instinctively. Within a year they were involved in medicine, agriculture, education, evangelism, and language translation. They adapted the work to their perception of needs.

My mother and father worked for seven years before anyone converted to Christianity, and, in fact, that first conversion came as a direct result of their compassion. Villagers would often abandon their sick outside our home, and my parents would care for them. Once, when a Hindu priest was dying of influenza, he sent his own sickly, nine-month-old daughter to be raised by my parents. None of his swamis would care for the frail child; he knew they would just let her die. My parents took her in, nursed her to health, and adopted her as their own. I gained a stepsister, Ruth, and my parents experienced an upsurge of trust from the villagers, some of whom embraced Christ's love for themselves.

Years later, when my widowed mother turned eighty-five, she helped forge a medical breakthrough. She had often treated gross abscesses on the legs of mountain people by draining the pus and excising a long, thin guinea worm. By studying the problem she learned that the worms spend their larval stage in water. From her familiarity with the villagers' habits, she concluded that wading in water was the likely means of transmission. As villagers waded in the water, guinea worms in their infected legs released larvae into the water supply, which the people then drank, perpetuating the cycle.

Having built up trust through decades of personal ministry, my mother rode her horse from village to village, urging the people to build stone walls around their shallow wells and to avoid foot contact with the water. People who had ignored government health workers listened to Granny Brand. Within a few years this elderly widow, a foreigner, single-handedly caused the eradication of all such worms, and their resulting abscesses, in two mountain ranges.

My wife, Margaret, had a similar experience with a condition afflicting the eyes of children. Whenever she encountered this condition, I could read it in the despair on her face that night. I would look at her and sympathetically murmur one word, "Keratomalacia?" and she would nod yes.

Keratomalacia results from a deficiency of vitamin A and protein among young children between one and two years old. A baby would thrive as long as it was breastfed, but soon a new brother or sister pushed it from its mother's breast. The new diet of rice failed to provide essential vitamins, making small bodies especially susceptible to infection. Finally, an outbreak of conjunctivitis, or pinkeye—an easily treatable infection for a well-nourished person—would attack the malnourished child's eyes. My wife, examining those eyes, saw a jellied mass, as if a strange heat ray had melted all the parts together. A glimpse of one of those children, fearfully squinting to keep out light, never failed to dishearten Margaret, regardless of how many successful procedures she had performed on other patients that day.

Spurred by Margaret's sense of urgency, some medical college re-searchers discovered that a common green herb, which grew wild all over our region, contained a remarkably high concentration of vitamin A. They also learned that peanuts, a local crop grown for oil, possessed the missing protein. Until then, the villagers had been feeding peanut residue to their pigs after mashing the nuts to produce oil.

Now the task became one of education. Margaret and public health nurses spread the word, and soon mothers were excitedly telling neighbors that the green herb and peanuts could prevent their children's blindness. The news traveled like gossip through the villages, protecting many children from the dreaded keratomalacia.

These two examples are hardly typical. Much of humanitarian work consists of exhausting labor with less dramatic results. Yet they demon-strate the importance of tactile Christian love. Although government health agencies and agricultural experts had sufficient knowledge to attack keratomalacia and the guinea worm, they had not gained the trust of villagers. A medical advance came, instead, from workers who were "in touch with" the suffering people and who had built up enough trust and respect to supply a remedy. I wonder how effective Granny Brand would have been had she dropped educational leaflets from a helicopter.

DR. PFAU'S LEGACY

One scene captures for me in a single image all the elements of the skin of Christ's Body. In the 1950s I visited a nun, Dr. Ruth Pfau, outside of Karachi, Pakistan, amid the worst human squalor I have ever come across. As the taxi neared her clinic, a putrid smell burned my nostrils, a smell you could almost lean on. Soon I saw an immense garbage dump by the sea, the city's accumulated refuse that had been stagnating and rotting for many months. The air was humming with flies.

At last I could make out human figures—people covered with sores—crawling over the mounds of garbage. They had leprosy, and more than a hundred of them, banished from Karachi, had set up home in this dump. Sheets of corrugated iron gave them a bit of shelter, and a single, dripping tap in the center of the dump provided their only source of water.

There, beside this awful place, I found Dr. Pfau's neat wooden clinic. She told me a bit of her life story: of the destruction of her childhood home in Leipzig by Allied bombers, of a scary time after the war when she walked at night with a teddy bear tucked under her arm to escape from Soviet-occupied East Germany, of her decision to convert after learning about forgiveness from a Dutch Christian who had survived Nazi concentration camps.

After training as a doctor, Pfau was sent to southern India by her order but ended up in Pakistan because of a visa issue. There, she visited a leprosy colony, where she met one of the million Pakistanis afflicted with the disease. She described the scene: "He must have been my age—I was at this time not yet thirty—and he crawled on hands and feet into this dispensary, acting as if this was quite normal, as if someone has to crawl there through that slime and dirt on hands and feet, like a dog."

The experience stunned her. "I could not believe that humans could live in such conditions," she said. "That one visit, the sights I saw during it, made me make a key life decision." That was when she moved to the little hut by the garbage dump to care for leprosy patients. A few years later she came to the hospital in Vellore to study our new surgical treatments for leprosy patients.

Throughout, Dr. Pfau lived in a single room, rising at 5 a.m. to pray and worship before tending to patients. When I visited her in Karachi, she proudly showed me her orderly shelves and files of meticulous records on each patient in the garbage dump. The stark contrast between the horrible scene outside and the oasis of love and concern inside her tidy clinic seared deep into my mind.

In the years after I met her, Dr. Pfau went on to establish 157 leprosy clinics across Pakistan. She became known as "Pakistan's Mother Teresa." Due to her efforts, in 1996 the World Health Organization declared Pakistan the first country in Asia to have controlled leprosy.

All over the world people like Dr. Pfau are fulfilling Christ's command to fill the earth with God's visible presence. They do so by exhibiting the properties of skin in the Body: beauty, sensitivity to needs, and the steady, fearless application of divine love through human touch.

CHAPTER TEN

BONE

A Necessary Frame

THE DANISH SETTING was worthy of a horror movie. Each morning I passed through a dark, narrow corridor and mounted creaking stairs that led to an ancient attic. There I found rows of boxes, layered with dust, filled with the moldering remains of six hundred skeletons. The rest of the day I hunkered over those boxes in the dimly lit room, sorting through bones. In all, I spent seven days crouched in the attic of that musty, old house in Copenhagen.

Shakespeare wrote, "The good [that men do] is oft interred with their bones." More than good is interred there. After a week I left that eerie place feeling as though I had watched a documentary on an ancient civilization. My only clues, tiny projections and furrows on the surfaces of bones exhumed from the dust of history, taught me much. Surfaces—skin, hair, and clothes, which consume so much human energy—had rotted away, leaving bones as the only mementos of those who had worn them.

The house served as a museum for Dr. Vilhelm Møller-Christensen, a medical historian, who invited me there because the skeletons had belonged to people with leprosy. After studying the bones, which he had discovered on an island off the coast of Denmark, Dr. Møller-Christensen wrote an extraordinary book on leprosy. Those of us who worked with the disease could hardly believe that he had learned so much without ever observing a living patient. All his insights came from poring over the five-hundred-year-old skeletons in his attic.

Picking over his clattery bones, much as a child rummages through a box of precious toys, Dr. Møller-Christensen would locate certain favorites and point out to me their features. "See this skeleton with missing front teeth," he said. "It demonstrates that leprosy first attacks the body's cooler parts." Together we examined the bones of feet and hands, speculating on what injuries might have caused their deformities.

I once heard a lecture by anthropologist Margaret Mead, who asked the question, "What is the earliest sign of civilization?" She suggested several possible answers. "A clay pot? Iron? Tools? Agriculture?" No, she said. "*This* is the evidence of the earliest true civilization," she declared as she held up a femur, a leg bone, that showed evidence of a healed fracture. Mead explained that the skeletal remains of competitive, savage societies never showed such signs of recovery. Clues of violence abound: ribs pierced by arrows, skulls crushed by clubs. But the healed femur shows that someone must have cared for the injured person—hunted on his behalf, brought him food, and served him at personal sacrifice.

Working alone in the attic one morning, I came across a large box of skeletons that showed such evidence of healing. I learned that Dr. Møller-Christensen had retrieved these particular bones from a monastery churchyard. An order of monks had ministered among leprosy patients, and now, half a millennium later, their compassion was manifest in the thin lines of healing where infected bone had cracked apart or eroded and then grown back together.

HIDDEN STRENGTH

TV crime shows such as *CSI* celebrate the feats of forensic scientists, who unravel the clues hidden in bones. Experts can determine a skeleton's age by how hard or ossified the cartilage has become. By age fifteen, the foot has fully formed, at twenty-five the collarbone has fused to the breastbone, and by age forty most of the seams in the skull have joined.

The bulky pelvis betrays the sex of the person who owned it. A broad and shallow pelvis with a smooth inner ring belongs to a woman, its oval opening precisely matching the size and shape that a baby's head needs to squeeze through. A man has a more narrow, heart-shaped pelvis, formed of heavier bones. The thickness of a bone may reveal more. Discus throwers and weight lifters have the densest bones because an exercised bone collects more calcium for needed strength. A horseback rider leaves clues in the stress lines of leg bones and pelvis; a furniture mover will show the effects in hip and shoulder bones.

No one has yet devised a material as well-suited for the body's needs. The only hard material in the body, bone possesses strength enough to support every other cell. Sometimes we press our bones together like a steel spring, as when a pole vaulter lands; other times we nearly pull a bone apart, as when lifting a heavy suitcase. In comparison, wood can withstand even less pulling tension and could not possibly bear the same compression forces. Steel, which can absorb both forces well, has three times the weight of bone and would limit us.

The economical body, using a weight-saving principle of architecture, hollows out this frame and fills the vacant space with a red blood cell factory that turns out seventeen million new cells in the time it takes to read this sentence. Bone sheathes life.

I find bone's design most impressive in the small, jewel-like chips of ivory in the foot. Twenty-six bones line up in each foot, about the same number as in a hand. Over the course of a match, a soccer player may subject these small bones to a cumulative force of a thousand tons. Not all of us leap and kick, but we do walk—on average, some sixty-five thousand miles, or more than two-and-a-half times around the world, in a lifetime. Reliable bone serves us without fanfare and grabs our attention only when we encounter a rude, fracturing force that exceeds its high tolerance.

Much of our planet's sedimentary rock consists of microscopic creatures that died, their skeletons cemented together to form rock. The ocean is a hungry place, and marine skeletons serve as much for protection as for movement. For mollusks, scallops, nautiluses, crabs, lobsters, and starfish, an external skeleton provides a suit of armor. The vast insect world also retains external skeletons, but these land-based species have size limits lest the burden of armor becomes insupportable.

On land, more subject to the incessant tug of gravity, movement is key to survival. The fastest rabbit evades the coyote, and the fleetest African cat dines on gazelle. An internal, *living* skeleton makes a dramatic improvement. An animal no longer needs to outgrow its home and risk a vulnerable molting period; rather, the skeleton now grows with the animal. And the design of muscles and ligaments attached to an internal scaffolding of bone allows heretofore unthinkable feats of movement.

Only with an internal skeleton can an animal the size of a condor support a ten-foot wingspan and soar on thermals for hours, or an elephant charge like thunder across the Serengeti, or a bull elk hoist his rack of antlers proudly toward the sky. Boneless locomotion tends to revert to the most primitive: the segmented scrunching of an earthworm or the lubricated slide of a slug.

Bones do not burden us; they free us.

THE FEATURE OF HARDNESS

Although no babies are born without bones, some inherit a genetic disorder called brittle bone disease. Their bones consist of deposits of calcium without the organic material of collagen that welds them together—the grit without the glue. A baby with this disorder may survive the pressures of birth, but with half its bones broken. Just diapering such a child may fracture a fragile hip or femur; a fall could break dozens of bones.

At our Carville hospital a patient taking steroids for treatment of her leprosy developed soft bones. She could break her foot bones by walking too briskly. Whenever I checked her x-rays for fractures, I was reminded that the most important feature of bone is its hardness. That property distinguishes it from all other tissue, and without it, bone is virtually useless.

Inside a spiritual Body, too, lives a core of truth that never changes: the frame supporting our relationships to God and to other people. Our age smiles kindly on musings about unity and diversity and sensitivity. We highlight other parts of the body—the heart on Valentine's Day, the face and skin in magazines and fashion, the hands in sculptures—and relegate the skeleton to Halloween, a spooky remnant of the past. I believe the quality of hardness deserves another look.

In some areas of my life I gladly accept restrictive rules. For instance, traffic laws inhibit my freedom—What if I don't want to stop at a red light?—yet I accept the inconvenience. I assume that some skilled engineers planned out the need for stoplights, and I prefer traffic laws to vehicular anarchy. Still, something within us rebels against the notion that someone else has determined how we should live.

In recent years, the democratic societies of the West have been enlarging the boundaries of acceptable behavior. The hookup culture recast sex, not as an expression of personal intimacy but as a way to experiment with multiple partners. Rock stars and college professors alike began advocating the use of hallucinogenic drugs. Pornography came out of the closet and grew into a multibillion-dollar industry. Binge-drinking swept across college campuses. Gender became fluid, more social construction than biology.

Those who opposed the new trends all too often came across as finger-wagging spoilsports. As it turned out, however, the traditions were right about many of those issues, at least in terms of their effect on physical health. The sexual revolution fostered the spread of venereal diseases, HIV/AIDS, and numerous other health problems. Now secular activists and public health officials are warning of the dangers associated with unprotected sex, binge-drinking, and drug use.

I once attended a conference that convened government agencies involved in health matters, including the Public Health Service, the Centers for Disease Control, the National Institutes of Health, and the Food and Drug Administration. Together, we set the goal of identifying the top ten health issues facing the United States.

I began jotting down the health concerns being discussed. It occurred to me that almost all of the primary health issues were lifestyle-related: heart disease and hypertension connected to stress; cancers associated with a toxic environment; AIDS contracted through drug use and sexual activity; sexually transmitted diseases; emphysema and lung cancer caused by cigarette smoking; fetal damage resulting from maternal alcohol and drug abuse; diabetes and other diet-related disorders; violent crime and automobile accidents involving alcohol or drugs. These were the endemic, even pandemic, concerns for health experts in the United States, and with the increase in opioid addiction and obesity the trend has only accelerated. Studies show that two-thirds of deaths prior to age sixty-five can be traced to behavioral choices.

I had attended comparable medical conferences in India, and the difference was striking. There, infectious diseases dominated the list of health concerns: malaria, polio, dysentery, tuberculosis, typhoid fever, leprosy. If I had suggested to Indian health experts the possibility of eradicating their top ten diseases, they could hardly imagine such a paradise. Yet look what has happened here. After conquering most of those infectious diseases, the United States has substituted new health problems for old, the majority of them stemming from lifestyle choices.

We have learned that what seems attractive and alluring may in fact prove damaging, and that some guidelines on behavior exist *for our own good*. As one researcher concluded, "In essence the studies empirically verify the wisdom of the book of Proverbs. Those who follow biblical values live longer, enjoy life more, and are less diseased." The state God desires for us, *shalom*, results in a person fully alive, functioning optimally to the Designer's specifications.

RESTRAINTS THAT FREE

He came to me as a patient in England: a burly Welshman who spoke with a workman's vocabulary. "Mornin', doctor," he growled. As he removed his wool, plaid jacket, I saw the reason for his appointment. The upper part of his right arm was not pink skin but grimy steel and leather—an awkward, brace-like contraption coated with black coal dust. I removed the brace, expecting to find evidence of a mining accident, and the puzzle deepened. His intact forearm led to a long section of flaccid flesh from elbow to shoulder. A section of bone appeared to be wholly missing.

After I studied the miner's records and x-rayed his arm, the answer fell into place. Years before, a bone tumor in his upper arm had led to a serious fracture, which splintered large pieces of bone. His doctor had deftly removed an eight-inch pipe of living bone and sewed back the tissue and skin around the space. As the miner lay recovering, his boneless upper arm seemed perfectly normal on the outside. Who would know the interior landscape had changed?

Everyone would know, the first moment this miner used the muscles still attached to remaining bone. Muscles work on a triangle principle, with a joint providing the fulcrum. To raise the hand, for example, the biceps muscle attached to the upper arm pulls up on the forearm. The arm bends at the elbow, completing the triangle. One muscle and one forearm bone do not make a triangle, however, and this coal miner lacked the third element, the humerus bone of the upper arm.

After his surgery, whenever the miner flexed his biceps muscle his entire upper arm shortened like a caterpillar contracting toward its middle. The space that should have housed a fixed bone between elbow and shoulder had become soft and collapsible, canceling out the triangle that would transfer force to his forearm. To compensate, the inventive Welsh doctor had fitted the miner with a crude exoskeleton, the leather and steel contraption. Now, when his biceps muscle contracted, these steel rods prevented his upper arm from shortening, thus allowing his forearm to lift upward.

I, too, have surgically removed humerus bones, though now we avoid the awkwardness of an external skeleton by inserting titanium or a bone graft into the vacant space. Nevertheless, this man's crude external brace had served him well for years, allowing him to work as a miner. He came to me asking for a new bone mainly because he had grown tired of having to buckle on his exoskeleton every day.

Because of its hardness, and susceptibility to fracture, bone can be seen as a constraint on human activity. It does, after all, keep us from squeezing into small spaces and from sleeping comfortably on hard ground. And what limits winter Olympians from adding twenty meters onto the graceful ski jump and keeps the slalom course in the domain of a few experts? The threat of broken bones. A person who breaks a leg in winter sports might wish for stronger bones, but stronger bones would be thicker and heavier, making skiing far more limited or impossible.

No, the 206 lengths of calcium that frame our body free rather than restrict us. Just as the Welsh miner's arm needed a proper scaffolding, whether external or internal, our ability to move effectively depends on bone—rigid, inflexible bone.

A DEPENDABLE SKELETON

I see a close parallel in the spiritual Body, where rules governing behavior function much like dependable bones. Moral law, the Ten Commandments, obedience—a "thou shalt not" negativism taints the words, and we tend to view them as opposites to freedom. As a young Christian, I cringed at such words. Later, though, especially after I became a father, I gave thought to the very nature of law.

I now see God's rules governing human behavior as guidelines intended to help us live the very best, most fulfilling life on earth. Rules may prove as liberating in social activity as bones are in physical activity.

Consider, for example, the Ten Commandments. The first four of those commandments set out rules governing a person's relationship to God: Have no other gods before me. Don't worship idols. Don't misuse my name. Remember the day set aside to worship me. As I contemplate these once-forbidding commands, more and more they sound like positive affirmations.

What if God had stated the same principles this way?

- I love you so much that I will give you *myself,* the only God you will ever need.

- I desire a direct relationship with you. Representations are inferior. You can have me.

- You will be known as "God's people" on the earth. Value the privilege; don't misuse it by profaning your new name or by not living up to it.

- I have given you a beautiful world to work in, play in, and enjoy. As you do so, set aside a day to remember where it came from. Your body needs the rest; your spirit needs the reminder.

The next six commandments govern personal relationships. One is already stated positively: honor your father and mother, a near-universal principle of family loyalty. The next five can be worded as follows:

- Human life is sacred, for human beings express my own image. You must honor the sanctity of life.

- Marriage can cure the essential loneliness in the human heart. Reserve physical intimacy for its rightful place within marriage lest you devalue and destroy that relationship.
- I am granting you a great privilege: the ownership of property. Stealing violates that right.
- I am a God of truth. A lie destroys contracts and promises, and undermines trust. You are worthy of trust: express it by not lying.
- I have given you good things to enjoy: cattle, grains, gold, music. Love people; use things. Do not use people for your love of things.

Restated positively, the commandments emerge as a basic skeleton of trust that links our relationship with God and with other people. By following them, we express something of God's image in us. God has given laws as the guide to the best life, whereas our innate rebellion tempts us to believe that God's laws are restricting us from something better.

HIGHER LAW

Yes, one might reply, perhaps the Ten Commandments can be manipulated with a positive spin. Why, then, didn't God state them that way? Why say, "You shall *not* murder. You shall *not* commit adultery. You shall *not* steal"?

I suggest two answers. First, a negative command actually proves less limiting than a positive one. "You may eat from any tree of the garden except one" allows more freedom than "You must eat from every tree of the garden, starting with the one in the northwest corner and working along the outer edge of the orchard." "You shall not commit adultery" is more freeing than "You must have sex with your spouse twice a week between the hours of nine and ten in the evening." "Do not covet" is less restrictive than "I am hereby prescribing limits on ownership. Every person is entitled to one cow, one ox, and three gold rings."

Second, humanity was not yet ready for an emphasis on the positive commands. The Ten Commandments represent a kindergarten phase of morality, setting forth the basic laws necessary for a community. When Jesus came to earth, he summarized the entire law in two positive commands: "Love the Lord your God with all your heart and with all your soul and with all your strength and with all your mind," and "Love your neighbor as yourself" (Luke 10:27). It is one thing not to covet my neighbor's property and not to steal from him, and quite another to love him so that I care for his family as much as I care for mine. Morality took a quantum leap from prohibition to love.

More, Jesus insisted that laws are given not for God's sake but for ours. "The Sabbath was made for man, not man for the Sabbath," he said (Mark 2:27). Elsewhere, "You will know the truth, and the truth *will set you free*" (John 8:32, emphasis added). Jesus came to cleanse the violence, greed, lust, and hurtful competition from within us *for our sakes*. He desires us to become more like God.

The Ten Commandments represent the fetal development of bone, the first ossification from cartilage. The law of love emerges as the fully developed, liberating skeleton. Hinged and jointed in the right places, it allows for smooth movement within the larger Body.

I have known people who feel compelled to cast off every possible limitation. They remind me of spoiled children, dashing from one toy to another, searching for yet another thrill, unaware that their search is actually a flight. Where do they stop cheating on their income tax? At what point do they allow the truth about an extramarital affair to leak out? At what lie will their friends or children cease to believe anything they say? Their lives become an entangling web of deception and fear. Does such a person have freedom?

We do not regard a skeleton as beautiful; rather, it contributes strength and functionality. I do not inspect my tibia and wish it to be longer or shorter or more jointed. I gratefully use it for walking, thinking about where I want to go rather than worrying about whether my legs will bear my weight. I should respond that way to the laws governing human nature. They establish the mere framework for relationships, which work best when founded on a few dependable principles of trust.

Of course, we can break the laws: adultery, thievery, lying, idolatry, and oppression of the poor have plagued every society in history. The resulting fracture brings to a halt the smooth functioning of the whole Body. Bones, intended to liberate us, enslave us when broken.

CHAPTER ELEVEN

HOW BONES GROW

WHEN I LIVED IN RURAL INDIA, we relied on walking as an essential mode of transportation. While tourists rode in automobiles and buses, missionaries and health workers who wanted to reach the village people traveled to roadless places. So they walked, and I viewed it as one of my most important jobs to get missionaries back on their feet after an accident.

One such missionary, Mrs. S., arrived at the hospital in Vellore and told me of an accident some months before in which she had broken her thigh bone, the femur. A local doctor in the mountains had set the bone, but his subsequent x-rays showed incomplete healing. Concerned, he sent her to our medical college.

After x-raying Mrs. S.'s fracture site, I expected to see the familiar sight of healing bone. The body's skeleton is a living, growing organ. When I cut bone, it bleeds. Most amazing of all, when bone breaks, it heals itself. Perhaps an engineer will someday develop a substance as strong and light and efficient as bone, but what engineer could devise a substance that can grow and also lubricate and repair itself?

A bone fracture sets an elaborate process into motion. Within two weeks a cartilage-like sheath surrounds the region, and mortar-laying cells then enter the jellied mass. These are the osteoblasts, the pothole-fillers of bone, which gradually replace the protective sheath with fresh bone. In two or three months a node of new bone marks the site, bulging over both sides of the broken ends like a spliced garden hose. Later, scouring cells will scavenge the surplus material so that the final result nearly matches the original bone.

To my surprise, I saw no evidence of this healing cycle in Mrs. S.'s x-rays. A clean line, an ominous gap, appeared between the two broken ends of bones, with no mending material fusing them together. I opened her leg for a firsthand look and could find no vestige of healing. Resorting to the inferior, nonliving tools of medicine, I fixed the area of the spiral break with a steel bone plate screwed into both pieces of the bone, above and below. Then I transplanted a grafted section of her tibia, to promote new bone formation, and sutured the wound.

After months of casts and wheelchairs and crutches, Mrs. S. again underwent x-rays. A milky cloud of growing bone showed the graft was taking, and yet a clean division between the sections of bone still yawned open. I knew we had something very unusual. After quizzing Mrs. S. and researching her history, I learned that twenty years earlier a doctor had irradiated her mid-thigh to treat a small, soft-tissue tumor. Besides the tumor, the radiation had apparently killed all her living bone cells at that site, and thus the two ends would never grow together.

The inactivity was driving Mrs. S. crazy. "God sent me to a place where I need legs!" she said. "We simply must find a solution."

So I performed another operation. I checked the steel plate. The two screws farthest from the fracture site were loose and easy to remove: her body had begun rejecting them. But the four screws nearest the fracture were as solid as if drilled into mahogany. I had to strain to turn them because the bone there was dead.

Obtaining two more bone grafts, one from Mrs. S.'s other tibia and one from her pelvis, I surrounded the fracture site with living bone, as if packing it in ice. Then I closed the wound and waited.

Mrs. S. recovered and rejoined her mission station in the mountains. She spent an active life trudging the dusty trails, and her improvised leg bone worked well enough. Seven years later, when she came in for a checkup, x-rays revealed that the original fracture site had never healed— in one sliver between the grafted bones I could see light. However, a living bone shell, like a knot on a tree, had joined the two pieces together and formed a misshapen bulge of bone. She walked almost normally thanks to the grafts—her original bone above, grafted bone in the middle, and her original bone below.

BONE IS ALIVE

Mrs. S. offered a rare example of dead and living bone tissue existing side by side. When I opened her leg, the two looked the same. They had one crucial difference: living bone interacted organically with her body while the dead bone did not. The dead bone had failed her, for only living bone can adapt to the needs of a living body.

The analogy from physical bone to a spiritual skeleton has already been drawn for us in a dramatic passage in Ezekiel 37. In a vision, the prophet tours a surrealistic valley piled high with dry bones. God directly addresses those bones: "I will attach tendons to you and make flesh come upon you and cover you with skin; I will put breath in you, and you will come to life. Then you will know that I am the LORD."

The bones in Ezekiel symbolized a nation, Israel, that had calcified into a dead skeletal form. Once vibrant, Israel's faith endured as a dry, lifeless memory. Yet in a vision of resurrection Ezekiel watched bones rattle together to form the framework for a new Body. A new nation would come to life with a preexisting heritage and a new understanding of God. The real value of a skeleton only comes to light when it supports a growing organism.

Bone is alive. The newborn baby has 350 bones that will gradually fuse together into the 206 carried by most adult humans. Many of the baby's bones are soft and pliable, lacking the quality of hardness, for the birth event would be impossible were they not so flexible. The same stages of growth that I saw in Mrs. S.'s graft site occur daily within the skeletons of children.

Observing the process of bone ossifying, I am reminded of my own skeleton of faith. As a newborn Christian my faith was soft and pliable, consisting of vaguely understood beliefs about God. Over time God has used the Bible and spiritual mentors to help ossify the framework of my faith. In the same way that osteoblasts lay down firm new minerals in a bone, the substance of my faith has become harder and more dependable. The Lord has become my Lord; doctrines that once seemed cold and formal have become an integral part of me.

Some believers act as if all answers can be codified in a comprehensive statement of faith. They treat others who doubt the basic doctrines as alien cells in the Body. Those of us tempted toward that kind of rigidity must come back to the analogy of living bone. New believers need time for the bones of faith to strengthen.

I have known many periods of doubt that tested my faith. In India I was challenged by exposure to other religions devoutly practiced by millions of people. In medical school I encountered professors who took for granted that the universe is based on randomness, with no place for an intelligent Designer. As I have grappled with these and other issues, I have learned the value of accepting as a rule of life something about which I have intellectual uncertainties. In other words, I have learned to trust the basic skeleton and rely on it even when I cannot figure out how various bones fit together and why some are shaped the way they are.

In medical school I studied under some of the pioneers of evolutionary theory, such as J. B. S. Haldane and H. H. Woolard. Some churches encouraged a kind of intellectual dishonesty on this subject. In the university their students took exams and recited the theories of evolution; when they joined the church, they declared their faith in a way that contradicted their exam answers. Ultimately this dichotomy led to a type of intellectual schizophrenia.

Only after much research and long periods of reflection was I able to put together what I had learned at church and what I had learned at school. In the meantime, I determined that my faith was based on realities that could stand on their own and did not need to be subordinated to an explanation from science. I operated with that assumption for years during which I was unable to resolve some of the mystery of how creation and evolution fit together. Indeed, in recent years, a new understanding of big bang cosmology and of the nature of DNA coding has greatly strengthened my faith in a guiding supernatural intelligence.

I have stood before a bridge in South America constructed of inter-locking vines that support a precariously swinging platform hundreds of feet above a river. I know that hundreds of people have trusted that bridge over the years, and as I stand at the edge of the chasm, I can see people confidently crossing it. The engineer in me wants to weigh all the factors—measure the stress tolerances of the vines, test any wood for termites, survey all the bridges in the area for one that might be stronger. I could spend a lifetime determining whether this bridge is fully trust-worthy. Eventually, though, if I really want to cross, I must take a step. When I put my weight on that bridge and walk across, even though my heart is pounding and my knees are shaking, I am declaring my position.

In my Christian walk I sometimes must proceed like this, making choices that involve uncertainty. If I wait for all the possible evidence, I'll never move.

AN EXOSKELETON'S STRANGE APPEAL

In the bayous of Louisiana where I live, twice a year a strange fever rolls through like a fog. Hand-painted signs appear outside dilapidated country restaurants: Fresh Crawfish Now! Schoolboys, barefoot and sweaty, scramble up the gullies dragging tin pails crawling with dozens of the prehistoric-looking creatures.

You can find crayfish (or crawfish) in almost any river, pond, or ditch in Louisiana. Menacing claws, half its body length, give the crayfish a militaristic appearance, like a gunboat with a couple of oversized how-itzers protruding over its bow. Two gleaming black eyes jut out between the claws, eyes that protrude on the ends of stalks—*movable* stalks. If the crayfish wants to see you from a better angle, he does not move his head but instead points his eyestalks in your direction. The rest of the crayfish duplicates in miniature its cousin the lobster: plates of overlapping armor ending in a broad, fan-shaped tail.

Crack open a crayfish and you'll find soft, white meat begging to be dipped in butter. No bones grow there to annoy a diner—the shell is its skeleton. In Louisiana, local restaurants will bring you platters of twenty or thirty of the creatures, their shells tinted bright red by the boiling process. After an hour of popping and scraping and digging, you leave a plateful of skeletons—thin, crayfish-shaped exteriors that, if propped up in a realistic pose, would pass for living crayfish.

A crayfish has an exoskeleton. Its muscles work against the carapace surrounding it, and that protective shell helps the crayfish survive in a ruthless world. Nevertheless, an exoskeleton presents certain disadvantages as well. A dog or cat or human being feels soft, warm, responsive. If you shake hands with a crayfish, you'll feel inflexibility, coldness, and probably pain. A good-sized lobster can break your finger with a quick pinch of its claw.

As I review the history of the Christian church, at times I see a basic misunderstanding of the place of the skeleton in the Body. Some believers wear their skeletons on the outside, and their dogma stands out as obtrusively as does a crayfish's shell.

Examples leap to mind, such as the monks known as "athletes for God." In the fifth century, Simon Stylites perched on a pillar east of Antioch for thirty-six years and is said to have touched his feet with his forehead more than 1,244 times in succession. Other monks subsisted by eating only grass. Theodore of Sykeon, a seventh-century saint, spent most of his life suspended from a rock in a narrow cage, exposed to the storms of winter, starving himself while soulfully singing psalms.

Some of these practitioners sought an extreme way to demonstrate their commitment to God. Others, however, made a public display of their zeal in order to impress onlookers—exactly the error that Jesus denounced in the Pharisees (see Matthew 23 and Luke 11).

More subtle forms of exoskeletons persist in Christian circles. Ask a non-Christian for her impressions of truly committed Christians, and she may well identify Christians by a list of things they avoid: smoking, drinking, swearing, gambling, tattoos, attending movies, dancing. As I hope I've made clear, it's *for our own good* to avoid certain harmful activities, but if that alone defines us, we have missed the positive, life-giving message of the gospel.

I am tempted to view legalism as a petty diversion. Does it matter if one denomination chooses to ban a harmless activity? Isn't it merely humorous that churches in some countries, whose members readily drink and smoke, recoil at the idea of Christians wearing blue jeans or chewing gum? Then I come across strong warnings in the New Testament. No other issue—not sexual sins, violence, or the behaviors which most rankle Christians today—inspired more fiery outbursts from Jesus than legalism.

Surprisingly, the people who most irritated Jesus were the Bible-belt fundamentalists of his day. This group, the Pharisees, gave away exact tithes, obeyed each minute law in the Old Testament, and sent out missionaries to gain new converts. Despite their upright behavior and the rarity of sexual sin or violent crime among the Pharisees, Jesus roundly censured these model citizens. Why?

To answer that question, I go back to the humble crayfish. In comparing its exoskeleton with my more advanced, internal skeleton, several differences shed light on Jesus' strong statements about the dangers of legalism.

First, the crayfish relies almost exclusively on its outside armor for protection. Humans, in contrast, have soft, vulnerable exteriors. Yet we form groups of the like-minded, and as the rules that define the group multiply and then calcify, we tend to hunker down inside them for protection. In effect, we develop an exoskeleton. In his *Letters to an American Lady*, C. S. Lewis wrote, "Nothing gives a more spuriously good conscience than keeping rules, even if there has been a total absence of real charity and faith."

Legalists fool us. Like the Pharisees and the "athletes for God," they impress onlookers with their fervor. Surely, you think, they have a high view of God. Growing up in a legalistic environment, though, I learned that legalism actually errs by *lowering* sights. It spells out exactly what a person must do in order to arrive at a state of moral superiority. In the process, legalists miss the whole point that grace is a gift freely given by God to people who don't deserve it.

Jesus told a story about two men praying in the temple. One, a righteous man, took pride in his superior moral character: "God, I thank you that I am not like other people—robbers, evildoers, adulterers—or even like this tax collector. I fast twice a week and give a tenth of all I get" (Luke 18:11-12).

The other man, a known sinner, could barely find the words to pray: "God, have mercy on me, a sinner." The Pharisee was comparing himself to people around him, whereas the tax collector compared himself only to God. The Pharisee rejoiced at being "holier than thou"; the tax collector recoiled at being "less holy than Thou." Jesus left no doubt as to which of the two God accepted.

In a further danger, legalism limits growth by forming a hard, crusty shell around the accepted group.

For an adult crayfish, growth can only occur once a year. In an arduous procedure known as *molting,* growth exposes the creature to peril as it sheds the confining exoskeleton. After several spasms of agitation, the crayfish pushes with all its might and its top plate of armor pops free. Gingerly, it removes its head, taking special care with the eyes and antennae. Finally, with a sudden spring forward, the crayfish unsheathes its abdomen and lies there, naked and exhausted.

As soon as it can, the crayfish slinks toward a rock or some other protection. Its body, once clad in lacquered chitin, now has the consistency of wet paper. During the next few weeks, the crayfish does all its growing for a year. It may add as much as an inch to its length before the new shell hardens into the shape and size of the new exoskeleton.

I have undergone a related process of faith molting. I started in a close-knit group that held rigid ideas of which Christians were worthy of fellowship. I tended to view the Body as an exclusive set of *people like me* encased in a shell. Inside, all was warm and comfortable; outside, the shell protected us from "the world." As I traveled and gained breadth of experience, I realized that not all Christians shared my assumptions about behavior, worship style, and doctrine. So I grew a new shell until the next experience came along.

In contrast, Jesus avoided language that might describe an exoskeleton. He kept pointing to higher, more lofty demands, using words such as *love* and *joy* and *fullness of life*—internal words. When someone came to him for a specific interpretation of a traditional law, he would point instead to the principle behind it. As he said to the Pharisees who criticized him for breaking Sabbath rules, "The Sabbath was made for man, not man for the Sabbath" (Mark 2:27). Jesus understood that the rules governing behavior are meant to free movement and promote growth as a vertebrate skeleton does, not to inhibit growth as an exoskeleton does.

A troubling phenomenon occurs among young Christians reared in solid homes and sound churches. After living their early years as models of faith, many become spiritual dropouts—as many as 50 percent, according to some surveys. Crayfish-like, did they develop a hard exterior that resembled everyone else's, only to find it restrictive and inadequate? When practiced mainly as an external exercise, faith can be cast aside in the manner of a crayfish flinging off its shell.

An outside shell may seem safe and attractive, and certainly it has advantages over no skeleton at all. God desires for us a more advanced type of skeleton, one that grows and adapts even as it remains hidden.

BLASTS AND CLASTS

I must turn to the microscope to see the living activity of bone. With enough magnification I can identify two types of active cells in bone. We have already met one type: the osteoblasts, pothole-filling repair cells that attach themselves to fracture sites and lay down bone crystal. The blast cells do not simply wait around for accidents. Billions of them patrol inside me on maintenance duty. In my youth the bustling blasts replaced 100 percent of the bone in my body each year. My jawbone as a four-year-old did not contain a single remnant of my three-year-old jaw bone; the shape stayed the same, only larger.

Bone does not surrender territory easily. It must be dynamited and vacuumed out, and for this job the body calls on a demolition team of oversized cells known as osteoclasts. The reckless clast cell leads a kamikaze life, boring through mineral with such verve that it burns out after forty-eight hours and is itself escorted away as waste.

Blasts adapt their design to the body's needs. If I break my foot and the pain causes me to adjust my walk so that I take shorter steps, the blasts will make alterations in my heel bone. If I take up weight-lifting, my bones will become thicker and develop extra struts. In fact, stress stimulates bone growth. Walking, lifting, flexing—any activity sends electrical currents through bone to generate growth.

Blasts predominate during the first half of a person's life, laying down new bone in the orderly scheme of growth. As I have aged, less than 20 percent of my bone gets replaced each year. Eventually, the demolition clasts will outstrip the weary restoration cells. With old age, teeth sockets decrease in size, the chin protrudes, the jaw angles in, and the elderly are often left with more severe, pointed faces. A common fracture may debilitate the elderly, for their blasts can barely manage the rigors of routine repair.

The skeleton of rules governing behavior in a spiritual Body must also adapt as it encounters new stresses. The basic principles, such as those set forth in the Ten Commandments and the Sermon on the Mount, do not change, but their specific application certainly changes, for many of the laws and observances of the Bible were geared to a society and culture alien to our own.

Consider the following list of direct instructions, all given to Christians in New Testament times. Some are still subscribed to by most Christians, while others are practiced only by members of a few denominations. I know of no community that obeys all of these instructions, which I've adapted from a list by the author Mont Smith:

1. Greet one another with a holy kiss (Romans 16:16).

2. Abstain from food sacrificed to idols (Acts 15:29).

3. Be baptized (Acts 2:38).

4. A woman ought to have a veil on her head (1 Corinthians 11:10).

5. Wash one another's feet (John 13:14).

6. It is disgraceful for a woman to speak in the church (1 Corinthians 14:35).

7. Sing psalms, hymns, and spiritual songs (Colossians 3:16).

8. Abstain from eating blood (Acts 15:29).

9. Observe the Lord's Supper (1 Corinthians 11:24).

10. Remember the poor (Galatians 2:10).

11. Anoint the sick with oil (James 5:14).

12. Permit no woman to teach men (1 Timothy 2:12).

13. Preach two by two (Mark 6:7).

14. Eat whatever is put before you without raising questions of conscience (1 Corinthians 10:27).

15. Prohibit women from wearing braided hair, gold, pearls, or expensive clothes (1 Timothy 2:9).

16. Abstain from sexual immorality (Acts 15:29).

17. Do not look for a wife (1 Corinthians 7:27).

18. Refrain from public prayer (Matthew 6:5-6).

19. Speak in tongues privately and prophesy publicly (1 Corinthians 14:5).

20. Lead a quiet life and work with your hands (1 Thessalonians 4:11).

21. Lift up holy hands in prayer (1 Timothy 2:8).

22. Give to those who beg from you (Matthew 5:42).

23. Only enroll (for aid) widows who are over sixty, have been faithful to their husbands, and are well-known for good deeds (1 Timothy 5:9-10).

24. Wives, submit to your husbands (Colossians 3:18).

25. Show no partiality toward the rich (James 2:1-7).

26. Owe no one anything (Romans 13:8).

27. Abstain from the meat of animals killed by strangulation (Acts 15:29).

28. If a man will not work, he shall not eat (2 Thessalonians 3:10).

29. Set aside money for the poor on the first day of every week (1 Corinthians 16:1-2).

30. If you owe taxes, pay taxes (Romans 13:7).

Biblical scholars can explain why the writer applied an underlying principle in just that particular way. For example, the apostle Paul gave instructions on eating meat that had passed through heathen ceremonies, hardly a problem in most nations today. And in a place like ancient

Corinth, women were judged by powerful social customs: if a woman spoke out in a public meeting, the group would naturally assume her to be a prostitute or pagan priestess.

Paul recognized the need to adapt lines of stress depending on the circumstances and the group. He refused to let Jewish Christians force Gentiles to be circumcised, yet he underwent purification rites in the Jerusalem temple (Acts 21) to win the trust of Jewish Christians.

Today we are facing new stress lines. As the population has multiplied and technology has increased, we need to place new emphasis on our responsibility to care for the planet. In a culture that objectifies sex, how can we reaffirm sex as a bond of intimacy and commitment and not as a haphazard expression of lust? My medical profession needs careful wisdom on new ethical issues. Today, we can prolong life almost indefinitely, even when the person has no consciousness or hope of recovery. Gene editing poses many new ethical questions by allowing us to manipulate genetic traits.

Although such issues do not call for sweeping revisions of creeds and beliefs, they do show the need for Christian leaders to reflect, study the Bible, and pray, and then help apply the will of God to their own generation. These teachers serve as living bone cells in Christ's Body, laying down the inorganic minerals that support our frame. They should possess humility and a commitment to preserve the basic principles of the faith, balanced by concern that the principles be relevant and give strength just where it is needed.

In 1892 Julius Wolff first noticed intersecting lines of stress in the cellular arrangement of the human skeleton, leading to Wolff's Law, which every medical student learns. Caught up in his enthusiasm, Wolff declared that bones were in a state of great flux, adapting readily to changes in environment and function. Actually, when I visit a museum, or a Copenhagen attic, and compare skeletons throughout the centuries, I am chiefly impressed by their uniformity. Adaptations to stress are minor knobs and slight ridges along bones that have maintained a consistent length and shape. The bone endures; the body adapts to new stresses.

PART FOUR

PROOF *of* LIFE

*A fly is a nobler creature than the sun, because
a fly hath life, and the sun hath not.*

St. Augustine

CHAPTER TWELVE

BLOOD
Life's Source

MY CAREER IN MEDICINE traces back to one dreary night at Connaught Hospital in East London.

Although my family had tried to influence me toward medicine, for a long time I stubbornly resisted all pressures to enter medical school. In truth, I was repulsed by the sight of blood and pus. Growing up in India, I shared in everything my parents did. Sometimes a patient came for treatment of an abscess, and when Dad dressed the wound, my sister and I held the bandages. My father had no anesthetics, so the patient would cling to a relative during the incision and drainage, and try not to cry out. Because of my vivid memories of those scenes and the sticky cleanup that followed, I dismissed any prospect of a career dealing with blood and pus.

Instead, I learned the building trade, apprenticing as a carpenter, a mason, a painter, and a bricklayer. I loved working with my hands and couldn't wait to return to India to practice my trade. In rural India, though, some knowledge of tropical medicine can prove vital, so the mission advised me to enroll in the same introductory course that my father had taken. I reported to Connaught Hospital to learn basic principles of diagnosis and treatment.

One evening during my stint there, my whole view of medicine—and of blood—permanently shifted. That night, hospital orderlies wheeled young accident victim into my ward. Loss of blood had given her skin an unearthly paleness, and her brownish hair seemed jet-black in contrast. Oxygen starvation had shut down her brain into a state of unconsciousness.

The hospital staff lurched into their controlled-panic response to a trauma patient. A nurse dashed down a corridor for a bottle of blood while a doctor fumbled with the transfusion apparatus. Another doctor, glancing at my white coat, thrust a blood pressure cuff at me. Fortunately, I had already learned to read pulse and blood pressure. I could not detect the faintest flicker of a pulse on the woman's cold, damp wrist. She did not seem to be breathing, and I felt sure she was dead.

In the glare of the hospital lights she looked like a waxwork Madonna or an alabaster saint from a cathedral. Even her lips were pallid, and as the doctor searched her chest with his stethoscope I noticed the blanched nipples on her small breasts. Only a few freckles stood out against the pallor.

The nurse arrived with a bottle of blood and buckled it into a metal stand as the doctor punctured the woman's vein with a large needle. They fastened the bottle high, using an extra-long tube, so that the increased pressure would push the blood into her body faster. "Keep watch!" the staff ordered as they scurried off for more blood.

Nothing in my memory can compare to the excitement of what happened next. The details of that scene come to me even now with a start. As the others all left, I nervously held the woman's wrist. Suddenly I could feel the faintest press of a pulse. Or was it my own finger's pulse? I searched again—it was there, a barely perceptible tremor. The next pint blood arrived and the staff quickly replaced the empty bottle. A spot ink appeared like a drop of watercolor on the patient's cheek and to spread into a lovely flush. Her lips darkened pink, then red, and ly quivered with a kind of sighing breath.

her eyelids fluttered lightly and parted. She squinted at first, and s contracted, reacting to the bright lights. At last she looked me. To my enormous surprise, she spoke. "Water," she said in ice.

woman entered my life for only an hour or so, and the me utterly changed. The memory of shed blood had kept me out of medicine; the power of shared blood ultimately led me to apply to medical school. I had seen a miracle, a corpse resurrected. If medicine, if blood could do this . . .

VITAL PIPELINE

Typically, blood gets our attention when we begin to lose it; the sight of it in tinted urine, a nosebleed, or a weeping wound provokes alarm. We miss the dramatic display of blood's power that I saw in the Connaught patient, the power that sustains our lives at every moment.

"What does my blood *do* all day?" I once heard a child ask, peering dubiously at his scraped knee. I turn to a technological metaphor to illustrate the answer. Imagine an enormous tube snaking southward from Canada through the Amazon delta, plunging into oceans only to surface at every continent—a pipeline so global and pervasive that it links every person worldwide. Inside that tube a plenitude of treasures floats along on rafts: produce from every continent, smartphones and other electronics, gems and minerals, all styles and sizes of clothing, the contents of entire shopping malls. Seven billion people have access: at a moment of need or want, they simply reach into the tube and take whatever product suits them. Somewhere far down the pipeline, a replacement is manufactured and inserted.

Such a pipeline exists inside each of us, servicing not seven billion but forty trillion cells in the human body. A renewable supply of oxygen, amino acids, salts and minerals, sugars, lipids, cholesterols, and hormones surges past our cells, carried on rafts of blood cells. In addition, that same pipeline ferries away refuse, exhaust gases, and worn-out chemicals. Five or six quarts of this all-purpose fluid suffice for all the body's cells.

Sixty thousand miles of blood vessels link every living cell. Highways narrow down to one-lane roads, then bike paths, then footpaths, until finally the red cell must bend sideways and edge through a capillary one-tenth the diameter of a human hair. In such narrow confines the cells are stripped of food and oxygen and loaded down with carbon dioxide and urea. From there, red cells rush to the kidneys for a thorough scrubbing, then back to the lungs for a refill. The express journey, even to the extremity of the big toe, lasts a mere thirty seconds.

A simple experiment reveals the composite nature of blood. Pour a quantity of red blood into any clear glass and wait. Horizontal bands of color will appear as various cells settle by weight until the final result resembles an exotic cocktail. The deepest reds, comprising clumps of red cells, sink to the bottom; plasma, a thin yellow fluid, fills the top part of the flask; white cells and platelets congregate in a pale gray band in between.

The body's survival depends on each of these cells. Platelets, for example—which have a delicate floral shape—play a crucial role in clotting. When a blood vessel is cut, the fluid that sustains life begins to leak away. In response, tiny platelets melt, like snowflakes, spinning out a gossamer web of fibrinogen. Red blood cells collect in this web, and soon the tenuous wall of red cells thickens enough to stanch the flow of blood. Platelets have a small margin of error. A clot too thick may block the flow of blood through the vein or artery and perhaps lead to a stroke. On the other hand, people whose blood has poor clotting ability live in constant peril: even a tooth extraction may prove fatal. A healthy body expertly gauges when a clot is large enough to stop the loss of blood yet not so large as to impede the flow within the vessel itself.

If any part of the network breaks down—the heart takes an unscheduled rest, a clot overgrows and blocks an artery, a defect diminishes the red cells' oxygen capacity—life ebbs away. The brain, CEO of the body, can survive intact only five minutes without replenishment.

Blood once repulsed me. Now, however, I feel like assembling all my blood cells and singing them a hymn of praise. The drama of resurrection enacted before my eyes in Connaught Hospital takes place without fanfare in each heartbeat of a healthy human being. Every cell in every body lives at the mercy of blood.

LIFE IN PERIL

To those of us who practice medicine, blood symbolizes life; that quality overshadows all other aspects. Every time I pick up a scalpel I have an almost reverent sense of the vital nature of blood.

In surgery I must control bleeding, for each quiver of the scalpel leaves a thin wake of blood. Most often it comes from a few of the millions of tiny capillaries, and I disregard them, knowing they will seal up of their own accord. Every minute or two a spurt of bright blood warns me of a nicked artery, which I must either clamp or sear with a cautery. The slow ooze of darker blood indicates a punctured vein, and I pay even closer attention. Having less muscle in its wall than an artery, a cut vein cannot easily close itself off. To avoid these problems, I try to locate each significant vessel before I make a cut, then I clamp it in two places and do my surgical work in the area between the clamps.

Despite all precautions, a different level of bleeding may occur—the surgeon's worst nightmare. Sometimes, through an error of judgment or loss of manual dexterity, a really large vessel gets cut or tears open and the wound gushes with blood. Welling up in the abdomen or the chest cavity, blood totally obscures the rip in the vessel from which it pours. The surgeon, as he fumbles in the sump of blood up to his wrists, shouts for suction and gauze sponges—and inevitably this is when the suction nozzle gets blocked or the lights go out. Few surgeons go through a career without such an incident.

I shall never forget the horror-struck face of one of my London students during one occurrence. He was performing a routine procedure on a woman in our outpatient clinic, excising a tiny lymphatic node from her neck for biopsy. The minor procedure required only a local anesthetic, which meant the patient was fully awake. I was working in an adjacent room when suddenly a nurse appeared in my doorway, her hands and uniform splashed with fresh blood. "Come quick!" she cried, and I dashed next door to find the intern, white as a corpse, working frantically on a woman from whose neck blood was gushing. It was hard to tell who was more terrified: my intern or the patient.

Fortunately, a masterful teacher had drilled into me the appropriate reflexes. I ran to the woman's side and, after removing all instruments from the wound, grasped her neck and applied firm pressure with my thumb. As my thumb filled in the broken part of the blood vessel wall, the bleeding stopped, and I stayed in that position until the woman calmed down enough for me to extend the anesthesia and repair the vessel. The intern had inadvertently snipped off a small section of the jugular vein!

My University College teacher, who bore the grand name Sir Launcelot Barrington-Ward and served as surgeon to England's royal family, had prepared us his students for just such an emergency. As his assistant, I would hear him ask each new student, "In case of massive bleeding, what is your most useful instrument?" At first the newcomer would propose various surgical tools, and the old teacher would frown and shake his head. There was only one acceptable answer: "Your thumb, sir." Why? The thumb is readily available—every doctor has one—and offers a perfect blend of strong pressure and gentle compliancy.

Then Sir Launcelot would ask, "What is your greatest enemy when there is bleeding?" and we would say, "Time, sir." And he would ask, "What is your greatest friend?" and we would say again, "Time."

While blood is being lost, time is the enemy. Second by second, life will leak away as the patient grows weaker. The surgeon must fight the temptation to panic, to grab at vessels and clamp them off with forceps here and there, often causing more damage.

Once I have my thumb on the bleeding point, however, time becomes my friend. Unhurried, I can pause and plan my next course of action. Meanwhile the body busies itself, forming clots to repair the breach. I can take time to clean up and arrange a transfusion or perhaps call for an extra assistant or enlarge the incision to get a better view. Once, I held a clump of blood vessels in my fist for twenty-five minutes while removing a diseased spleen, operating with one hand while I dammed the flow of blood with the other. All this can happen if my thumb is pressing firmly on the area of bleeding. And when I finally remove my thumb, my assistants poised to act, I usually find that no action is needed. The bleeding has stopped.

At those moments, in the rush of adrenaline brought on by the crisis, I have a sense of spiritual exaltation. I feel at one with the millions of living cells in that wound fighting for survival. I realize, with a sense of humility and awe, that a common thumb is the only thing preventing my patient's death.

After many such experiences in the electric atmosphere of the operating room, every surgeon learns to identify blood with life. The two are inseparable: you lose one, you lose both.

Why, then, does blood as a Christian symbol seem to contradict what I learn at such moments?

A TOAST TO LIFE

Although modern worshipers may feel uncomfortable with the fact, the Christian faith is inescapably blood-based. Old Testament writers spell out the details of blood sacrifices, and their New Testament counterparts overlay those rituals with theological interpretations. And daily, weekly, monthly (or whenever, depending on denomination), we are called upon to commemorate Christ's death with a ceremony centered in his blood.

I admit at the outset that I sometimes find the associations of the blood symbol distasteful. I switch on my radio on a Sunday morning while driving from my hospital in Carville to New Orleans. A Southern pastor is leading a Communion service in a church in the bayous. The congregation murmurs as he holds up a four-inch thorn and illustrates how barbarously the soldiers jammed a crown of them onto Jesus' head. He describes the scene of a cross being strapped to a back bloodied by whips. Every occasion for the word *blood*—the nailing, the thud of the cross in the ground, the spear in the side—seems to give this preacher a fresh burst of energy.

I drive along in bright Louisiana sunshine, glancing outside at the stately egrets, white as clouds, bobbing for food in the canals lining the highway. As if in mocking contrast, the theme of death spills from the car radio. The preacher asks his parishioners to think of their recent sins, one by one, and to contemplate the horrible guilt that led to such a bloody death on their behalf. A ceremony follows, the sacrament itself.

My mind, jarred from the solemn church service, returns to the literal substance of blood—not the watery purple liquid filling Communion cups but the rich, scarlet soup of proteins and cells that keeps my patients alive. Again I wonder: Has something been lost over the centuries, something foundational? The Louisiana pastor focuses on *shed* blood—but does not the sacrament center also on blood that is *shared*?

Medically, blood signifies life and not death. Blood feeds and sustains every cell in the body with its precious nutrients. When it seeps away, life falters. Has our modern use of the symbol, as illustrated by the radio preacher's fixation on death, strayed so far from the original meaning?

Deep in the biblical record lies a fundamental association of blood with life. In a covenant with Noah, God commanded, "You must not eat meat that has its lifeblood still in it" (Genesis 9:4). Later, in the formal legal code given to Moses and the Israelites, God reiterated the command as "a lasting ordinance for the generations to come." Why? "Because the blood is the life, and you must not eat the life with the meat" (Leviticus 3:17; 7:26-27; 17:11, 14; Deuteronomy 12:23).

Old Testament Jews felt no squeamishness about blood, and in that sheep-and-cattle culture everyone witnessed the bloody deaths of animals. Even so, every good housewife checked her meat to see that no blood remained. The rule was absolute: do not eat the blood, for it contains life. Kosher cuisine developed elaborate techniques to ensure that no blood contaminated the meat.

In view of this background, consider the shocking, almost revolting message Jesus proclaimed to that culture:

> Very truly I tell you, unless you eat the flesh of the Son of Man and drink his blood, you have no life in you. Whoever eats my flesh and drinks my blood has eternal life, and I will raise them up at the last day. For my flesh is real food and my blood is real drink. Whoever eats my flesh and drinks my blood remains in me, and I in them. Just as the living Father sent me and I live because of the Father, so the one who feeds on me will live because of me. (John 6:53-57)

Those words, coming at the peak of Jesus' popularity, signaled a turning point in his public acceptance. The Jewish audience became so confused and outraged that a crowd of thousands, who had pursued Jesus around a lake in order to crown him king, silently stole away. Many of his closest disciples deserted him; enemies plotted to kill him. Jesus had gone too far.

Jesus spoke as he did not to offend but rather to effect a radical transformation in the symbol. God had said to Noah, if you drink the blood of a lamb, the life of the lamb enters you—don't do it. Jesus said, in effect, if you drink my blood, my life will enter you—do it! For this reason, I believe Jesus intended our ceremony to include not only remembrance of his past death but also realization of his present life. We cannot sustain a spiritual life without the nourishment his life provides.

The ceremony we call Eucharist (or Lord's Supper or Holy Communion or Mass) has its origin in Jesus' last night with the disciples before his crucifixion. There, amid a stuffy roomful of his frightened disciples, Jesus first said the words that have been repeated millions of times: "This is my blood of the covenant, which is poured out for many for the forgiveness of sins" (Matthew 26:28). Jesus commanded his disciples to drink the wine, representing his blood. The offering was not merely poured out but rather taken in, ingested. "Drink from it, all of you" (v. 27).

That same evening Jesus used another metaphor, perhaps to underscore the meaning of shared blood. "I am the vine; you are the branches," he declared. "If you remain in me and I in you, you will bear much fruit; apart from me you can do nothing" (John 15:5—echoing the wording of John 6:56). Surrounded by the vineyard-covered hillsides that ringed Jerusalem, the disciples could more easily comprehend this metaphor. A grape branch disconnected from the nutrients of the vine becomes withered, dry, dead, useless for anything except kindling. Only when connected to the vine can the branch bear fruit.

Even in the doom-shrouded atmosphere of that last night, at the meal from which the sacrament derives, the image of life wells up. For the disciples, the wine would symbolize Jesus' blood, which could vitalize them much as the sap does the grapevine.

If I read these accounts correctly, they correspond to my medical experience precisely. I do not believe that blood represents life to the surgeon but death to the Christian. Rather, we come to the table also to partake of his life. "For my flesh is real food and my blood is real drink. Whoever eats my flesh and drinks my blood remains in me, and I in him"—at last those words make sense (John 6:55-56). Christ came not just to give us an example of a way of life but to give us life itself. Spiritual life is not ethereal and outside us, something that we must work hard to obtain; it is in us, pervading us, like the blood that flows through every living body.

Theologian Oscar Cullmann, in *Early Christian Worship*, presents a fresh interpretation of Jesus' first miracle, when he turned water into wine at a wedding banquet in Cana. Cullmann says this miracle or "sign," points to Jesus' new covenant, linking the wedding wine to the wine of the Last Supper. The setting could hardly be more appropriate to introduce this great symbol: a wedding feast, filled with joyous music, the laughter of guests, the clink of pottery, the sounds of celebration—so different in tone from the dreary sounds I heard on my radio in Louisiana.

The very institution that recalls the death of Jesus is also a toast, if you will, to the Life that conquered death and is now offered freely to each of us.

TRANSFUSION

The history of blood transfusion, like that of so many other medical techniques, began perilously and quickly sped toward disaster. In 1492, the same year Columbus set sail, a doctor in Italy tried transfusing blood from three young boys into ailing Pope Innocent VIII. All three donors died of hemorrhaging, while the thrice-punctured pontiff barely outlived them.

Not until the nineteenth century did medicine achieve some success with the mysterious procedure of transfusion. In England, Dr. James Blundell saved the lives of eleven of fifteen women who were hemorrhaging after childbirth. Extant etchings record the scene: a solemn Blundell looks on as a healthy volunteer, standing, delivers her blood through a tube directly into the vein of a dying woman. Those etchings capture the human-to-human essence of shared life that today gets lost in the formality of computer-matched blood banks and sterile containers.

Even so, for many years blood transfusions involved great risk. Decades of bewilderment passed before researchers sorted out the complexities of blood typing, Rh factors, and the prevention of clotting. During World War I, the benefits of blood transfusion finally began to outweigh its dangers. Word spread rapidly among the troops: "There's a bloke who pumps blood into you and brings you back to life even after you're dead!"

The dramatic experience of watching a blood transfusion at London's Connaught Hospital had drawn me into medicine. Twelve years later, with medical and surgical training behind me, I found myself back in the land of my birth, India. I arrived as an orthopedic surgeon at the Christian Medical College in Vellore, which was recruiting specialists from all over the world. These included Dr. Reeve Betts from Boston, who went on to become the father of thoracic surgery for all of India.

When Betts first arrived, he ran up against an immediate roadblock—the lack of a blood bank. Betts had the experience and skill to save the lives of patients who began streaming to Vellore from long distances, but he could do nothing without blood. Thus, in 1949, a blood bank became my number one priority. I had to learn the skills needed for typing, cross-matching, and screening donors for health problems. In India, where so many people were afflicted with parasites or a hidden virus of hepatitis, we struggled to make our system foolproof.

The attitudes of Indian people themselves offered the biggest challenge. They instinctively understood that blood is life, and who can tolerate the thought of giving up lifeblood? Their resistance dearly tried Betts's patience. "How could anyone refuse to give blood that would save his own child?" he would mutter darkly after emerging from a lengthy family council.

In a typical case, Betts scheduled surgery on a twelve-year-old girl with a diseased lung. First, he informed the family that the girl would die unless he removed the lung. The family members nodded with appropriate gravity. Betts continued, "The surgery will require at least three pints of blood, and we have only one, so we need you in the family to donate two more." At that news, the family elders huddled together, then announced a willingness to pay for the additional pints.

I watched Betts flush red. The veins in his neck began to bulge, and his shining bald head made an excellent barometer of his remaining tolerance. Working to control his voice, he explained that we had no other source of blood—it could not be purchased. They might as well take the girl home and let her die. Back to the conference. After more lively discussion the elders emerged with a great concession. They pushed forward a frail old woman weighing perhaps ninety-five pounds, the smallest and weakest member of the tribe. "The family has chosen her as the donor," they reported.

Betts fixed a stare on the sleek, well-fed men who had made the decision, and then his anger boiled over. In halting Tamil he blasted the dozen cowering family members. Although few could understand his American accent, everyone caught the force of his torrent of words as he jabbed his finger back and forth from the husky men to the frail woman.

Abruptly, with a flourish Betts rolled up his own sleeve and called over to me, "Come on, Paul—I can't stand this! I won't risk that poor girl's life just because these cowardly fellows can't make up their minds. Bring the needle and bottle and take my blood." The family fell silent and watched in awe as I dutifully fastened a cuff around Betts's upper arm, swabbed the skin, and plunged the needle into his vein. A rich, red geyser spurted into the bottle and a solemn "Ahh!" rustled through the spectators.

At once there was a babel of voices. "Look, the sahib doctor is giving his own life!" Onlookers called out shame on the family. I reinforced the drama by warning Betts not to give too much this time because he had given blood last week and the week before. "You will be too weak to perform the surgery!" I cried.

In this case, as in most others, the family got the message. Before the bottle was half full, two or three came forward and I stopped Betts's donation to take instead their trembling, outstretched arms. In time, his reputation spread throughout the hospital: if a family refused blood, the great doctor himself would contribute.

OLD SYMBOL, NEW MEANING

In a time when blood transfusion was unknown, Jesus chose the perplexing image of drinking his blood. Who can describe the process by which Christ's body and blood become a part of my own? Jesus used the analogy of branches attached to a vine, and the more contemporary metaphor of blood transfusion helps me to grasp a deeper symbolism.

My experiences with blood transfusion, beginning with that night at Connaught Hospital, underscore the life-giving power of blood. The Communion service reminds me that Christ is not dead and removed from me, but alive and present within me. Every cell in his Body is linked, unified, and bathed by the nutrients of a common source. Blood feeds life.

Thus, the Lord's Supper has become for me, not an embarrassing relic from primitive religion but an image of startling freshness. I can celebrate the sensation of coming to life through the symbol of Christ's blood transfused into me. The woman at Connaught Hospital escaped death because of the shared resources of a nameless donor; Dr. Reeve Betts's patients gained new hope through the contributions of individual family members; similarly, I receive in the Eucharist an infusion of strength and energy by availing myself of Christ's own reserves.

Under the old covenant, worshipers brought the sacrifice—they gave. In the new, believers *receive* tokens of the finished work of the risen Christ. "My body, which was broken *for you* . . . my blood which was shed *for you* . . ." In those phrases, Jesus spans the distance from Jerusalem to me, cutting across the years that separate his time from mine.

When we come to the table we come short of breath, with a weakened pulse. We live in a world far from God, and during the week we catch ourselves doubting. We muddle along with our weaknesses, our repeated failings, our stubborn habits, our aches and pains. In that condition, bruised and pale, we are beckoned by Christ to his table to celebrate life. We experience the gracious flow of God's forgiveness and love and healing—a murmur to us that we are accepted and made alive, transfused.

"I am the Living One," Christ said to the awestruck apostle John in a vision. "I was dead, and now look, I am alive for ever and ever!" (Revelation 1:18). The Lord's Supper sums up all three tenses: the life that was and died for us, the life that is and lives in us, and the life that will be. Christ is no mere example of living; he is life itself. No other New Testament image expresses the concept of "Christ in you" as well as blood. As George Herbert writes,

> Love is that liquor sweet and most divine,
> Which my God feels as blood; but I, as wine.

CHAPTER THIRTEEN

WISE BLOOD

I TURN UP THE COLLAR OF MY WOOL TOPCOAT and bow my
head against the moisture-laden wind. Snow is gradually transforming
the modern city of London into a Dickensian Christmas card, floating
down to cover potholes, gutter, cars, and sidewalk with a blanket of softly
glowing white. From somewhere I hear music, the muffled tones of brass
and what seems like human voices. On a night like this?

I walk toward the sound, the music growing louder with each step,
until I round a corner and see its source: a Salvation Army band. A man
and a woman are playing a trombone and trumpet, and I wince as I
imagine metal pressed against lips in the numbing wind. Three other Sal-
vationists are lustily singing a hymn based on a poem by William Cowper.

> There is a fountain filled with blood
> Drawn from Emmanuel's veins;
> And sinners, plunged beneath that flood,
> Lose all their guilty stains.

An unavoidable smile crosses my face as I hear those words. I have just
come from hospital rounds where real blood was being drawn from some
veins, transfused into others, and diligently scrubbed off surgical smocks
and nurses' uniforms. With my church background, I understand the
origin and meaning of the Christian symbol. But for the bystanders, what
images must fill their minds as they hear that hymn? Would not the

phrase "washed in the blood of the Lamb" seem to the modern Briton as bizarre as a report of animal sacrifice in Papua New Guinea?

Nothing in modern culture corresponds to the notion of blood as a cleansing agent. We use water, with soap or detergent, to clean. Blood, we try to scrub *off*, not scrub with. What possible meaning could the hymn writer, and Bible writers before him, have intended?

In fact, modern medical science reveals that the jarring symbol of cleansing conforms closely to blood's actual function. The image immortalized in Cowper's hymn reflects good biology as well as good theology.

If you truly wish to grasp the function of blood as a cleansing agent, I suggest a simple experiment. Find a blood pressure kit and wrap the cuff around your upper arm. Have a friend pump it up to about 200 mm. of mercury, a sufficient pressure to stop the flow of blood in your arm. Initially, your arm will feel an uncomfortable tightness beneath the cuff. Now comes the revealing part of the experiment: Perform any easy task with your cuffed arm. Flex your fingers and make a fist several times in succession, or cut paper with scissors, or drive a nail into wood with a hammer.

The first few movements seem quite normal as the muscles obediently contract and relax. Then you feel a slight weakness. Almost without warning, after perhaps ten movements, a hot flash of pain strikes. Your muscles cramp violently. If you force yourself to continue, you will likely cry out in absolute agony. Finally, you give up, overcome by pain.

When at last you release the tourniquet and air escapes from the cuff with a hiss, blood rushes into your aching arm and a wonderfully soothing sense of relief floods your muscles. The pain is almost worth enduring just to feel that bracing recovery. Your muscles move freely, soreness vanishes. Physiologically, you have experienced the cleansing power of blood.

The pain came because you forced your muscles to keep working while the blood supply to your arm was shut off. As muscles converted oxygen into energy, they produced certain waste products (metabolites) that normally would have been flushed away by the bloodstream. Because of the constricted blood flow, however, these metabolites accumulated in your cells. They were not *cleansed* by a steady stream of blood, and so in a few minutes you felt the agony of retained toxins.

No cell lies more than a hair's breadth from a blood capillary, lest these poisonous byproducts pile up. The bloodstream flowing inside these narrow capillaries simultaneously releases its cargo of fresh oxygen and absorbs hazardous waste products, which it transports to the kidneys. There, the renal artery divides and subdivides into a tracery of a million crystal loops so intricate that some observers judge the kidneys second in complexity only to the brain. After the kidney has sorted through the blood's payload to extract some thirty chemicals, it promptly reinserts the rest back into the bloodstream. One second later, the thunder of the heart resounds and fresh blood surges in to fill the kidney's tubules.

One group of people view the kidney with an attitude approaching reverence: those unfortunate enough to have dysfunctional kidneys. Fifty years ago all of them would have died. Now they have time to contemplate the wonders of the kidney—too much time. Thrice a week for five hours they lie or sit still while a tube drains all their blood through a noisy, clanging machine the size of a large suitcase. This technological marvel, a kidney dialysis machine, crudely replaces the intricate work of the soft, bean-shaped human kidney. Our natural one, however, weighs only one pound, works around the clock, and normally repairs itself. Just to be safe, our body provides a spare—one kidney would do the job just fine.

Other organs join the scavenging process. A durable red cell can only sustain the rough sequence of freight-loading and unloading for a half million circuits or so until, battered and leaky as a worn-out river barge, it nudges its way to the liver and spleen for one last unloading. This time, the red cell itself is picked clean, deconstructed into amino acids and bile pigments for recycling. The tiny core of iron, a "magnet" for the crucial hemoglobin molecule, gets escorted back to the bone marrow for reincarnation in another red cell. A new cycle of fueling and cleansing begins.

SPIRITUAL CLEANSING

Medically, blood sustains life by carrying away the chemical byproducts that would interfere with bodily processes—in short, by cleansing. As I reflect on a spiritual Body, the blood metaphor offers a fresh perspective on a perpetual problem in that Body: sin. To some, the word *sin* has become fusty and timeworn. Blood, however, provides the perfect analog to reveal the nature of sin and forgiveness with bright clarity. Just as blood cleanses the body of harmful metabolites, forgiveness cleanses away the waste products—sins—that impede true health.

Too often we think of sin as a private list of grievances that happen to irk God, and in the Old Testament God seems easily irritated. Yet even a casual reading of the Old Testament shows that sin is a blockage, a paralyzing toxin that inhibits our full humanity. To repeat a lesson learned from the Ten Commandments, God gave laws for *our* sakes. In the midst of a withering attack on Israel recorded in Jeremiah, God makes this poignant observation: "They pour out drink offerings to other gods to arouse my anger. But am I the one they are provoking? . . . Are they not rather harming themselves, to their own shame?" (Jeremiah 7:18-19).

Pride, egotism, lust, and covetousness work like poisons to interfere with healthy relationships. Sin results in separation—from God, other people, and my true self. The more I cling to my private desires, my thirst for success, and my own satisfactions at the expense of others, the further I will drift from God and others.

The Israelites had a stark pictorial representation of this state of separation between God and humanity. God's presence rested in a Most Holy Place, approachable only once a year (the Day of Atonement) by one man, the high priest, who had purified himself through an elaborate ritual. Jesus Christ made that ceremony obsolete by his once-for-all sacrifice. "This is my blood of the covenant, which is poured out for many *for the forgiveness of sins,*" he said as he instituted the Last Supper (Matthew 26:28, emphasis added).

The Lord's Supper, or Eucharist, as celebrated today, contrasts sharply with the Day of Atonement. No longer must we approach God through a ritually purified high priest. On the day Jesus died, the thick temple veil of separation split from top to bottom. Now all of us can enter into direct communion with God: "We have confidence to enter the Most Holy Place by the blood of Jesus, by a new and living way opened for us through the curtain, that is, his body" (Hebrews 10:19-20).

The Lord's Supper commemorates Christ's sacrifice as continuous and ongoing. Wine stands as a symbol for blood that both bathes every cell with the nutrients of life and also carries away accumulated waste and refuse. Following our analogy, in the act of repentance each cell willingly avails itself of the cleansing action of blood. Repentance is for our sakes, not to punish us but to free us from the harmful effects of accumulated toxins. "This is Christ's body, broken *for you*"—for *your* gossiping, *your* lust, *your* pride, *your* insensitivity—broken to remove all those and to replace them with new life.

Why do any of us go to church and sit on rather uncomfortable furniture, lined up in rows as in a school classroom? Perhaps because each of us feels a pang of longing and hope—a hope to be known, to be forgiven, to be healed, to be loved. Something like this longing lies at the heart of the ceremony of the Lord's Supper.

Symbols are weaker than the reality behind them. Christ has given us the wine and the bread, received in person and ingested, as signs that we are forgiven, healed, and loved. The elements work their way inside us, becoming material as well as spiritual nourishment, bearing a message to individual cells throughout each body.

If sin is the great separator, Christ is the great reconciler. "For if, while we were God's enemies, we were reconciled to him through the death of his Son, how much more, having been reconciled, shall we be saved through his life!" (Romans 5:10). He dissolves the membrane of separation that grows up every day between ourselves and others, between ourselves and God. "But now in Christ Jesus," said Paul, "you who once were far away have been brought near by the blood of Christ. For he himself is our peace" (Ephesians 2:13-14).

Near the end of his life, François Mauriac, the French Catholic novelist who received the Nobel Prize for Literature, reflected on his love-hate history with the church. He detailed the ways in which the church has not kept its promise: the rifts and compromises and moral failures that have always characterized it. The church, he concluded, has strayed far from the precepts and example of its founder. And yet, added Mauriac, despite all its failings the church has at least remembered two words of Christ: "Your sins are forgiven you," and "This is my body broken for you." The Lord's Supper brings together those two words in a quiet ceremony of healing.

OVERCOMING

Blood has one more property that gives rich meaning to a puzzling biblical image. In one of his visions, the apostle John describes a cosmic confrontation between the forces of good and of evil. He describes the ultimate victors over the personification of evil this way: "They overcame him by the blood of the Lamb" (Revelation 12:11 NKJV).

How can such a word apply to blood? *Overcome* connotes strength and power: a terrorist with a knife or gun overcomes an airplane crew; a Japanese sumo wrestler overcomes his opponent. On the other hand, blood connotes weakness and failure—a bleeding person has *been* overcome.

How to make sense of this strange combination of words? Once again, the biology of blood hints at an answer. To grasp the meaning, I search for ways in which physical blood might be said to *overcome*. The relatively recent procedure of immunization reveals the process.

When the body confronts a new invader, it normally requires hours to break the code and manufacture antibodies to combat the threat. For centuries humanity lived at the mercy of this time gap, which sometimes resulted in the annihilation of entire populations. Immunization, which derives from brilliant pioneering work by Edward Jenner and Louis Pasteur, ingeniously solved the time problem. By exposing the body to a weakened or "killed" virus in advance, a vaccine gives the body a crucial advantage, shrinking the time gap. Now the body can flood the scene with prepared antibodies and quickly overwhelm the intruders.

No other medical procedure has done more to conquer disease, as the story of smallpox vaccination wonderfully demonstrates. From the dawn of civilization smallpox ravaged the world, killing half a billion people in the nineteenth century and almost that many in the twentieth. After a stupendous effort of vaccination, in 1980 the World Health Organization declared it the first infectious disease in history to be eradicated.

On a visit to Bogotá, Colombia, I stood at the foot of a bronze memorial to twenty-two young boys who became living links in a courageous saga of immunization. Their story brings to light the secret of how blood overcomes.

In the year 1802, a smallpox epidemic broke out among the natives and Spanish settlers of Bogotá. Aware that the disease could easily decimate an unprotected population, the city's ruling council petitioned Spain's King Carlos IV for help. The king had an active interest in the new technique of vaccination, for his three children had received Jenner's controversial treatment. But how could cowpox vaccine be transported to the New World? Within Europe, vaccinators ran threads or quills through cowpox sores and stored them in glass vials for delivery to other countries, but the virus would dry up long before a ship could cross the Atlantic.

One of the king's advisers suggested a daring and innovative plan. An expedition was born, with the grandiloquent name Real Expedición Maritima de La Vacuna (Royal Maritime Expedition of the Vaccine), headed by the physician Francisco de Balmis. Soon the Spanish ship *Maria Pita* left port with a cargo of twenty-two boys, aged three to nine, conscripted from a nearby poorhouse. De Balmis vaccinated five boys before departure; the others would form a human chain to keep the virus alive.

Five days into the voyage, vesicles—small craters with raised edges and sunken centers—appeared on the infected boys' arms, and on the tenth day lymph flowed freely from the mattering sores. De Balmis carefully scraped that valuable lymph into the scratched arms of two uninfected boys. Every ten days two new boys were selected, vaccinated with the live virus, and quarantined until their harvesting time.

By the time the *Maria Pita* reached Venezuela, the very last boy was keeping the vaccine alive, the sole hope for staving off further epidemics. De Balmis selected twenty-eight more boys from the local population and stayed long enough to vaccinate twelve thousand people. DeBalmis's assistant headed for the original destination, Bogotá, now in desperate straits due to the long delay. A moment of panic struck when his ship wrecked on the way, but the carriers of the live vaccine survived. Everyone in Bogotá was vaccinated, smallpox soon faded away, and the assistant went on to vaccinate Peru and Argentina.

Meanwhile, de Balmis himself journeyed to Mexico, where he launched a frenzied vaccination campaign. After crisscrossing the country, he organized a new boatload of volunteers for the dangerous voyage to the Philippines. Those islands, too, eventually received protection from the unbroken human chain that stretched all the way back to an orphanage in La Coruña, Spain. Hundreds of thousands of people lived in debt to those original twenty-two orphans.

HOW BLOOD OVERCOMES

As a child in India I experienced the full force of person-to-person transmission. My parents had very limited quantities of vaccine and no facilities for cold storage, so they relied on the same source as did de Balmis: previously vaccinated human beings. Runners would bring the vaccine up mountain paths and hand the precious lymph to my father. Even before the runner caught his breath, Father would break the little tubes of lymph and begin vaccinating the waiting crowd. Later, from one infected arm he would draw enough lymph to vaccinate ten other villagers. Those ten yielded enough to vaccinate a hundred more. The blood of each vaccinated person locked away the memory of the pox virus, which allowed it to counter any threat posed by smallpox.

From this property of blood *overcoming* through person-to-person transmission, I gain new insight into the biblical use of the word. For example, at a very tender moment, during his last evening with the disciples before his crucifixion, Jesus said this: "In this world you will have trouble. But take heart! I have overcome the world" (John 16:33). In view of what followed, that declaration has a hollow tone, and those triumphant words must have soured for the disciples as they cowered in the darkness watching Jesus' death on a cross.

Later, in the book of Revelation the image of a Lamb appears again and again to represent Christ. We can easily miss the irony of the weakest, most helpless animal symbolizing the Lord of the universe—and not only that but a lamb "looking as if it had been slain" (Revelation 5:6). This, then, provides the background for the phrase that used to puzzle me: "overcome by the blood of the Lamb."

A pattern emerges: God responds to evil not by obliterating it but by making evil itself serve a higher good. Jesus overcame evil by absorbing it, taking it on himself, and, finally, by forgiving it. Jesus overcame as the One who goes before, by going right through the center of temptation, evil, and death.

Think of a scientist staring through her microscope at a microbe population that threatens the world. She longs for a way to remove her lab coat, shrink down to micron size, and enter that microbe world with the genetic material needed to correct it. Now imagine God, after observing with great sadness the virus of evil that has infected creation, joining humanity in order to vaccinate us against its effects. An analogy points to truth weakly; nothing could have more force than the simple assertion, "He became sin for us."

The author of Hebrews spells out what God's Son accomplished:

> Since the children have flesh and blood, he too shared in their humanity so that by his death he might break the power of him who holds the power of death—that is, the devil—and free those who all their lives were held in slavery by their fear of death. For surely it is not angels he helps, but Abraham's descendants. For this reason he had to be made like them, fully human in every way, in order that he might become a merciful and faithful high priest in service to God, and that he might make atonement for the sins of the people. Because he himself suffered when he was tempted, he is able to help those who are being tempted. (Hebrews 2:14-18)

Somehow, by drawing on the resources of Christ, I become better equipped to overcome my own temptations.

When we lived in Vellore, an epidemic of measles struck the city, and one of my daughters came down with a severe infection. We knew she would recover, but our infant daughter Estelle was much more vulnerable because of her age. When the pediatrician explained our need for convalescent serum, word went around Vellore that the Brands needed the "blood of an overcomer."

While not actually using those words, we called for someone who had contracted measles and had defeated that disease. We located such a person, withdrew some of his blood, let the cells settle out, and injected the convalescent serum. Equipped with borrowed antibodies, our daughter successfully fought off the disease. She overcame measles not by her own resistance or vitality but as a result of a battle that had taken place previously within someone else.

A person's blood becomes more potent as that person prevails against outside invaders. After antibodies have locked away the secret of defeating each disease, a second infection of the same type will normally do no harm. A protected person has "wise blood," to use a term Flannery O'Connor originated. Recall the just-quoted passage from Hebrews: "Because he himself suffered when he was tempted, he is able to help those who are being tempted" (Hebrews 2:18). And again, "We do not have a high priest who is unable to empathize with our weaknesses, but we have one who has been tempted in every way, just as we are—yet he did not sin" (Hebrews 4:15).

Today, when we partake of Communion wine, we are reminded to reflect on the wise and powerful blood of the One who has gone before. It is as though our Lord is saying to us, "This is my blood, which has been strengthened and prepared for you. This is my life which was lived for you and can now be shared by you. I was tired, frustrated, tempted, abandoned; tomorrow you may feel tired, frustrated, tempted, or abandoned. When you do, you may use my strength and share my spirit. I have overcome the world for you."

CHAPTER FOURTEEN

BREATH

Inspiration and Expiration

WHAT TREE RIVALS THE BANYAN in extravagance? It extends roots
not only from its trunk but also from its branches, dozens and eventually
hundreds of sinewy stalks wending toward the ground to sprout root
systems of their own. Uninterrupted, a banyan tree will grow forever,
renewing itself in the radial extremities even as the inner core dies of old
age. A single tree may cover acres of land, becoming a self-perpetuating
forest spacious enough to shelter a full-scale bazaar (its name comes
from the Hindi word *banian:* a caste of merchants).

A majestic specimen thrives in Calcutta (Kolkata) today, preserved in
the city's botanical gardens. The Great Banyan Tree looks like a giant,
bushy tent supported by colonnades of wooden poles. Somewhere in the
midst of that thicket of roots and branches, the central trunk began
growing two centuries ago. After sustaining damage from a fungus and a
cyclone, the inner core was removed in 1925, yet the outer tree grows on.

For a child who likes climbing and swinging from vines, the banyan
tree provides endless amusement. As a six-year-old, I explored one for
several days when my parents camped under a banyan during a missionary venture. As they went about their medical and spiritual work, my
sister and I played Swiss Family Robinson inside the huge tree. The stalks
that fell like stalactites from upper branches made for ideal climbing.

Even better, some helpful visitors had looped and knotted various vines to form high swings and trapezes.

I was swinging on one such loop through a corridor within the tree, calling for my sister to push me higher and higher. As the height of my arc increased, I felt another loop of tendril brush against the back of my neck on my back-swing. I ducked to avoid it but failed to duck again as the swing lurched forward. The high loop caught under my chin, causing the swing to abruptly stop. The vine had clamped shut my windpipe, and I could not speak or scream or even breathe. I hung suspended in air, like a puppet tangled in its wires. My sister, on the ground, made a few frantic efforts to pull me loose and then must have run for help.

I woke up sometime later with my mother stooping over my camp cot, pleading with me to speak to her. When I said "Mother," she burst into tears. She had feared brain damage, and my first word came as a welcome relief.

Other than a sore neck and a slight skin abrasion resembling a rope burn, I bore no lasting damage from the experience, though for many years I harbored a primitive terror of breathlessness. Anything that covered my mouth and nose, even immersion in water, brought back that terror, and I would fight as if for my life. I learned that lack of breath does not feel like anesthesia or sleep; it feels like death.

FUEL OF LIFE

Since my experience in the banyan tree, I have seen many medical situations that corroborate the terror I felt that day. Emergencies of all types produce panic—heart attack victims clutch their chests; people with brain damage may thrash violently; soldiers in war stare, unbelieving, at a severed limb—but I know of no human experience that produces an uncontrolled panic to rival breathlessness.

The marathon runner staggers across the finish line with mouth agape, her ribs heaving, her head bobbing like a rooster's, her whole body jerking in a bellows motion until gradually the oxygen floods in and the emergency subsides. The runner herself, however, feels no panic, for she has planned to end the race in a critical oxygen debt. Her gestures underplay those of a person who *must* have oxygen: eyes bulge, hands grasp frantically at empty air, and the heart races. Oxygen shortage sets in motion a vicious cycle, for the accelerated heartbeat, trying to distribute faster what little oxygen is present, requires even more oxygen.

We live, all of us, five or six minutes from death. Existence depends on our access to oxygen, the fuel that keeps our vital fires burning. When deprived of air, a person actually turns blue, first around the fingernails, tongue, and lips—a projection of the internal drama onto the visible screen of skin. High school biology students learn what causes the color shift: blue blood lacks the supply of oxygen from the lungs that normally turns it a rich scarlet.

The earth supplies an atmosphere we humans can relate to, and if our physical bodies leave the earth, for example on a space mission, we must somehow reproduce the oxygen in that atmosphere. Indeed, the entire animal kingdom subsists on this one life-sustaining element. Some animals' means of collecting oxygen show remarkable beauty: the jewel-like fronds of marine worms, the fluted gills of tropical fish, the brilliant orange skirt of a flame scallop. Our own lungs opt for function, not form, yet they work well enough to make an engineer sigh in envy.

I remember being a medical student first cutting into a corpse's torso. I had studied the various internal organs: heart, kidneys, liver, pancreas. But the lungs! Supremely important, they crowd all the rest, spilling into every crevice and cranny. Pump in air to simulate breathing and they expand as if to burst out of the chest cavity. Such vital organs need the space.

On an average day our lungs expand and contract around seventeen thousand times, ventilating enough air to fill a medium-size room or blow up several thousand party balloons. Any slight change in effort, such as climbing stairs or running for a bus, can double the demand for oxygen, and an involuntary switch orders a speeded-up rate of breathing. Receptors scattered around the body constantly monitor oxygen and carbon dioxide to determine the ideal rate.

Breathing proceeds without conscious control during sleep, or we would die. And the utilitarian body borrows the same air-flow system for such acts as speaking, singing, laughing, sighing, and whistling. My own love affair with breath, which began after my young body hung from the banyan tree, only intensified after I studied the mechanisms involved.

I have watched many patients live out the drama of breathing. Shortly after arriving in India as a physician, I received two telephone calls on the same day, one from Calcutta and one from London. Both concerned the medical predicament of a young polo player. The only son of a wealthy British lord, he had come to Calcutta to learn international finance on behalf of his father's global network of banks. The doctors in Calcutta and his relatives in England urged me to take the very next flight to Calcutta to examine the young man who, the day after a strenuous polo match, had become suddenly paralyzed with polio.

Over a staticky phone line I shouted instructions for the hospital to prepare an iron lung and also to perform a tracheostomy if he developed breathing difficulties. Then I dashed to the Madras (Chennai) airport to catch the night flight. When I arrived in Calcutta, a car sped me to his hospital bedside.

I have a lasting impression of the figure I found inside that hospital room. A life of good nutrition and much leisure time on the rugby and polo fields had given the patient a superb physique. His arm and leg muscles, though paralyzed, bulged even in repose. He had built up an enormous lung capacity, now virtually useless apart from the assistance of an iron lung. The machine worked on a bellows principle, mechanically pushing his chest in and out to replace what his lungs could no longer do on their own.

The cruel irony of the scene struck me: that marvelous body shoved inside an ugly metal cylinder that noisily forced air in and out of his lungs. I thought briefly of Michelangelo's sculptures, *The Captives,* in which magnificent bodies seem trapped in marble despite their efforts to free themselves. Before me, this man's athletic body lay enclosed in steel. Nurses told me he had felt flu-like symptoms on Friday but had gone ahead with a polo match on Saturday so as not to disappoint his teammates. At the onset of polio, exertion may prove deadly.

To my dismay I learned that the hospital had not performed the tracheostomy, so I immediately sent for an anesthetist. I worried about the fluid that collects when muscles used for coughing or clearing the throat no longer function. I explained to the young athlete what we planned to do. An aide who stood beside him assured me that money was no object and we should take any measures that might help.

The young man himself responded, in two sentences. He could only say one word per breath, and that with great effort. Every sound came in a clicking, wheezing, almost choking expulsion of air. "Give...me...breath," he said and paused. I leaned closer to hear him over the rhythmic pumping of the iron lung. And then, "What...is...the...use...of...money... if...you...can't...breathe?" I looked at his face with great sadness.

After reassuring him we would do all we could, I stationed a nurse with a throat suction by him and went downstairs for a cup of coffee and bite of breakfast. The anesthetist had not yet arrived and I, having missed a full night of sleep, sought a little nourishment to improve my concentration. I had not finished my coffee when a nurse came with the news that the patient had died. Evidently he had regurgitated some fluid, which obstructed the flow of oxygen, and the suction device could not keep up. His breathing stopped, and with it his life.

WIND, BREATH, SPIRIT

The English language describes breathing as a succession of two acts: inspiration and expiration. "I have expired" means I have breathed out; "I have inspired" means I have breathed in. If changed slightly to "I am inspired," it could mean I am filled with an enlivening breath from the artistic muses or, in a religious context, filled with the Holy Spirit. The writers of the Bible claim to have been inspired or in-breathed.

Our vocabulary, superb and precise when describing the material world, falters before the inner processes of spirit. The very word *spirit* in many languages means nothing more than breath or wind. Thus Greek and Hebrew use exactly the same word for the Spirit of God, biological breathing, and even the wind gusts from a storm.

The words *spirit* and *wind* or *breath* have a clear affinity, as Jesus' conversation with Nicodemus shows: "The wind blows wherever it pleases. You hear its sound, but you cannot tell where it comes from or where it is going. So it is with everyone born of the Spirit" (John 3:8). An invisible force from far away, whether wind or Spirit, has visible manifestations. And as a dying person breathes his or her very last breath and *expires,* life departs. Breath becomes air. Although the body remains intact, breath and spirit leave hand in hand.

Philosophers and theologians have written books exploring these and other links, but I will limit my comments to an aspect of breathing I deal with daily. I must stay close to what impressed me first as a six-year-old dangling limply from a tree, and then again as a doctor watching my patients' last few expirations. Breath sustains life. Any interruption in the fuel it supplies causes immediate death (the fastest poisons, such as curare and cyanide, work by interrupting the transport and absorption of oxygen).

Analogously, our entire faith begins here. We are told in so many words that eternal life cannot consist of mere oxygen and other nutrients. For eternal life we must establish a connection to a different kind of environment. Jesus makes it clear: "Very truly I tell you, no one can enter the kingdom of God unless they are born of water and the Spirit. Flesh gives birth to flesh, but the *Spirit gives birth to spirit*" (John 3:5-6, emphasis added). I think of an astronaut on the moon, or someday Mars, who must rely on an oxygen source to survive. Spiritual life will likewise fail unless we have contact with a spirit like the wind, the Holy Spirit.

"Blessed are those who hunger and thirst for righteousness," said Jesus—a picture comes to my mind of a runner gasping for breath or an athlete in an iron lung—"for they will be filled" (Matthew 5:6). The psalmist called up the image of a deer panting for streams of water; "so my soul pants for you, my God," he cried (Psalm 42:1). God's Spirit offers the only adequate solution to the spiritual "oxygen debt" hinted at in these phrases.

I confess a hesitance to write on spirit. Has any aspect of faith become more muddled? The word *spirit* itself, taken from a metaphor as common as the air outside, remains nebulous and imprecise. As one trained in scientific disciplines, I find it much easier to write of the material world that I can touch and see and analyze. And yet there is no Christian faith without spirit. Because God is spirit, only spirit can convey the image of God in us, Christ's Body on earth.

Present in the original act of creation, the Spirit of God hovered over the waters as matter came into existence. The Spirit inspired the prophets through the spiritual droughts and famines of Old Testament history. The Spirit anointed Jesus at the beginning of his ministry and was passed on to the disciples when Jesus *breathed* on them (John 20:22). We need the Spirit, said Jesus, for the quickening—a new birth, as Jesus once termed it—necessary to enter the kingdom of God.

At Pentecost the Holy Spirit (with "a sound like the blowing of a violent *wind*") entered and dramatically transformed a tiny band that was to become the church. This event more than anything caused church leaders to include the Spirit as a separate person within the Godhead. They could not exclude the Spirit: evidence seemed as real and convincing as evidence for another Person whom they had seen and touched.

The Holy Spirit, then, allows the reality of God's own self to establish a presence inside each one of us. God is timeless, but the Spirit becomes for us the present-tense application of God's nature—the Go-Between God, in Bishop John Taylor's lovely phrase. Correspondence with the Spirit keeps us spiritually alive.

I CAN BREATHE!

Alas, the life of the spirit is neither as instinctive nor as urgent as the mode of breathing in the physical body. We can get out of breath spiritually and yet not sense it. Breath may choke off slowly, unnoticed at first until a constant state of energy shortage sets in. I saw the physical parallel to this spiritual process in a woman I treated in London.

She came to me as a patient during my medical residency—a middle-aged widow and a hard worker—complaining about her recent tendency to drop things. "My hands tremble," she said, "and just this week two of my best china cups broke when they slipped out of my fingers. I get so tired, and now I can't seem to control my hands or my nervous disposition."

"I must be getting old," she concluded with a deep sigh and a shaky voice. I told her that fifty was certainly not old and that I would try to locate the physical cause for her condition. As she described a variety of symptoms, I began to suspect thyrotoxicosis, a disease of the thyroid gland that can cause tremors and shakiness.

First I felt for a thyroid swelling but found none. When a chest x-ray showed a shadow behind the upper end of her breast bone, I examined her neck again, this time probing with my fingers down into the base of her neck while she swallowed. Indeed there was an obstruction—I felt a rounded lump rise up out of her chest and touch my fingers. Her windpipe also seemed bent over to one side.

Another x-ray of her upper chest revealed that the rounded shadow had compressed her windpipe. I asked, "Do you have trouble breathing?"

"No, not at all," she replied, to my surprise. "I just get tired."

I explained that I believed her problem stemmed from a lump of cells that had grown in an unusual place in her thyroid gland, causing thyrotoxicosis. The lump had further extended into her chest, and because of the possibility of cancer, we needed to remove it. Otherwise, she might soon find it difficult to take a breath.

I assisted at the surgery conducted by my chief. We started at the neck, fully prepared to saw through bone to open up her upper chest if necessary. After some gentle pulling, the lump popped into view. It was fibrous and well-nourished, the size of an orange. It had indeed bent the windpipe, constricting it from both sides. We removed the tumor and closed the wound.

I next saw the woman a few weeks later when she returned for a checkup. She rushed up to me, and even before I had a chance to greet her, she almost shouted, "I can breathe!"

I was puzzled. "Were you afraid the operation would stop your breathing?" I asked.

"No, no, you don't understand," she said with great excitement. "Now I can *breathe* for the first time in years and years. I can run up stairs! I feel like a teenager again. I can breathe!"

Her condition became clear. That lump must have been growing slowly for fifteen years or more, gradually compressing her trachea, like a boa constrictor tightening its grip. The woman had adapted, at first by stopping frequently to catch her breath. It bothered her, but since she knew elderly people who became breathless and unable to climb stairs, she assumed she too had an aging heart. Over time she learned to walk very slowly and mount steps one at a time. In her eyes she had become prematurely old, and the hand tremors corroborated this doddery image of herself.

Now, however, she could take great gulps of air and dash upstairs. Over fifteen years she had forgotten how good it felt to breathe deeply and freely. I was struck by the near-miraculous change in the posture, facial expression, and total attitude toward life of this woman with the retrosternal thyroid lump. Absolute ecstasy spread across her face as she swelled her chest and announced loudly, "I can breathe!"

SPIRITUAL FUEL

Occasionally I try to savor the pleasure of God's good gifts, like breathing, by imagining for a moment that I have lost them. I hold my breath and pretend my trachea is blocked. I sense the rising panic spreading throughout my body. I envision my red corpuscles turning blue. I hear a drumming in my head. Then suddenly I open my mouth and suck in a gulp of air. I blow out carbon dioxide and vapor and then distend my chest and let the air rush in. I feel a short burst of the relief and ecstasy experienced by the woman with thyrotoxicosis.

The cells of my body need the fuel of oxygen to survive. Herbert Spencer expressed the scientific principle: whatever amount of power an organism expends in any form equals the power that was taken into it from without. The same principle holds true in the spiritual world. Christ's Body needs breath, the inspiration of his Spirit. One epistle warns, "Do not quench the Spirit"—or "put out the Spirit's fire," as some versions have it (1 Thessalonians 5:19). We need the stream of life that comes from God, and only the Spirit can provide that.

The Old Testament gives one striking example of spiritual renewal as performed by a Jewish official in the worldly government of Babylon. For Daniel, praying toward Jerusalem meant a public display of civil disobedience punishable by imprisonment or death. Disregarding the king's edict, three times a day Daniel flung open his window and turned toward Jerusalem for his prayers. Surely when he did so, the reality of Babylon—the aroma of spices and produce from the bazaar, the strange language, the jumble of urban noise—blew in. Yet as Daniel prayed in the midst of that foreign culture, he also breathed in a kind of spiritual oxygen that reminded him of a different reality.

Daniel's practice of facing Jerusalem stemmed from a prayer articulated by Solomon at the dedication of the temple when he asked God to hear any call for help prayed toward the temple in Jerusalem. Jesus later discounted the notion that location mattered in worship (see his dialogue with a Samaritan woman in John 4), and most of us do not pray toward a geographical place. Still, the scene captures for me the concept of planting my feet firmly on earth while sighting along a line of spiritual direction. I need a time of day to orient myself, to bring heaven and earth together. In the midst of the clamor and tumult of this material world, I must find a place of quietness to listen to the still, small voice for guidance of my life.

I too live in an alien culture that bombards me with false values of lust, pride, violence, selfishness, and materialism. To survive, I must pause to breathe in the power of the living God and consciously direct my mind to how God wants me to live. Spiritually, I cannot survive the foreign atmosphere of earth without live contact through the Spirit. Daniel looked out over the streets of Babylon, but his mind and soul were in Jerusalem. The astronauts walked in the cold, forbidding atmosphere of the moon only by carrying with them resources from another world to keep them alive. I need just that kind of daily reliance on the Spirit of God.

CHAPTER FIFTEEN

BODY *in* MOTION

A FRAIL MAN with a more-than-prominent nose and a face seamed with wrinkles crosses the stage. His shoulders slump slightly and his eyes seem sunken and cloudy; he has, after all, passed his ninetieth birthday. He sits on a stark black bench, adjusting it slightly. After a deep breath, he raises his hands, which, trembling slightly, hang poised for a moment above a black and white keyboard. Then the music begins. All hints of age and frailty slip away from the minds of the four thousand people gathered to hear Arthur Rubinstein.

He has chosen a simple program: Schubert's Impromptus, several of Rachmaninoff's Preludes, and Beethoven's familiar *Moonlight Sonata,* any of which could be heard at a music school recital. They could not, however, be heard as played by Rubinstein. Defying mortality, he weds a flawless technique to a poetic style, producing music that provokes prolonged shouts of "Bravo!" from the wildly cheering audience. Rubinstein bows slightly, folds those marvelous nonagenarian hands, and pads offstage.

A bravura performance such as Rubinstein's engrosses my eyes as much as my ears. Hands are my profession, and I have studied them all my life. To me, a piano performance is a ballet of fingers, a glorious flourish of ligaments and joints, tendons, nerves, and muscles. I must sit near the stage to watch their movements.

I know that some of the passages, such as the powerful arpeggios in *Moonlight's* third movement, require responses too fast for the pianist to accomplish consciously. Nerve impulses do not travel fast enough for the brain to sort out that the third finger has just lifted in time to order the fourth finger to strike the next key. Months of practice pattern the brain to treat the motions as subconscious reflex actions—"finger memory" musicians call it.

I marvel too at the slow, lilting passages. Rubinstein controls his fingers independently, so that when striking a two-handed chord of eight notes, each of the fingers exerts a slightly different pressure, with the melody note ringing loudest. A few grams more or less pressure in a crucial pianissimo passage has such a minuscule effect that only a sophisticated laboratory could measure it. The human ear contains just such a laboratory, and musicians like Rubinstein gain acclaim because discriminating listeners can savor those subtle nuances of control.

Often I have stood before a group of medical students or surgeons to demonstrate the motion of one finger. I hold before them a dissected cadaver hand, almost obscene-looking when severed from the body and trailing strands of sinew. I announce that I will move the tip of the little finger. To do so, I must place the cadaver hand on a table and spend perhaps four minutes sorting through the intricate network of tissues. Finally, when I have arranged at least a dozen parts in the correct configuration, with care I can maneuver them so the little finger bends without buckling the proximal joints.

In order to allow dexterity and slimness for actions such as piano playing, the finger contains no muscles; tendons transfer force from muscles in the forearm and palm. In all, seventy separate muscles contribute to hand movements.

I could fill a bookcase with surgery manuals suggesting various ways to repair hands that have been injured. Indeed, I have written one myself: *Clinical Mechanics of the Hand*. Yet in forty years of study I have never read a technique that succeeds in improving a normal, healthy hand. Computer scientists have developed programs that can defeat grand masters at chess, but the most sophisticated robots cannot come close to duplicating the fluid motions of a four-year-old at play.

I remember my lectures as I sit in concert halls watching slender fingers glissade across the keyboard. I revere the hand; Rubinstein took its function for granted. Often he closed his eyes or gazed straight ahead and did not even watch his hands. He was not thinking about his little finger, he was contemplating Beethoven and Rachmaninoff.

Scores of other muscles lined up as willing reinforcements for Rubinstein's hands. His upper arms stayed tense, and his elbows bent at nearly a ninety-degree angle to match the keyboard height. Shoulder muscles rippling across his back contracted to hold his upper arms in place, and muscles in his neck and chest stabilized his shoulders. When he came to a particularly strenuous portion of music, his entire torso and his leg muscles went rigid, forming a firm base to allow the arms leverage. Without these anchoring muscles, Rubinstein would topple over every time he shifted forward to touch the keyboard.

I have visited facilities that produce radioactive materials. With great pride, scientists show off their expensive machines that allow them to avoid exposure to radiation. By adjusting knobs and levers they can control an artificial hand whose wrist supinates and revolves. More advanced models even possess an opposable thumb, an advanced feature reserved for primates in nature. Smiling like a proud father, the scientist wiggles the mechanical thumb for me. I nod approvingly and compliment him. But he knows, as I do, that compared to a human thumb his atomic-age hand is clumsy and primitive, a child's Play-Doh sculpture compared to a Michelangelo masterpiece. A Rubinstein concert proves that.

LEARNING TO MOVE

Six hundred muscles, composing 40 percent of our weight (twice as much as bones), burn up much of our energy in order to produce motion. Tiny muscles govern the light permitted into the eye. Muscles barely an inch long allow for a spectrum of subtle expressions in the face—a bridge partner or a diplomat learns to read them as important signals. A much larger muscle, the diaphragm, controls coughing, breathing, sneezing, laughing, and sighing. Massive muscles in the buttocks and thighs equip the body for a lifetime of walking. Without muscles, bones would collapse in a heap, joints would slip apart, and movement would cease.

Human muscles are divided into three types: smooth muscles control the automatic processes that rumble along without our conscious attention; striated muscles allow voluntary movements, such as piano-playing; and cardiac muscles are specialized enough to merit their own category. A hummingbird heart weighs a fraction of an ounce and beats eight hundred times a minute; a blue whale's heart weighs half a ton, beats only ten times per minute, and can be heard two miles away. In contrast to either, the human heart seems dully functional, yet it does its job, beating 100,000 times a day with no time off for rest, to get most of us through seventy years or more.

Muscles pack enough potential to allow the Bolshoi Ballet and the graceful sports of ice skating and gymnastics. On TV the performers seem models of weightless beauty, gliding through the air, pirouetting on a single toe, dismounting from a high bar with a light spring. Up close and in person, though, motion seems more like hard work. It is *noisy* there, all shocks and thuds and creaking boards and panting, sweating bodies. That humans can transform such strenuous activity into fluidity and grace testifies to the dual nature of motion: robust strength and masterful control.

The movements of a Rubinstein or a Michael Jordan do not come cheaply. The motor cortex of the brain, on which will be written all the coding for intentional movement, starts out as blank as a washed chalk-board. An infant, dominated by gravity, cannot hold her head or trunk upright. Her hand and leg movements are abrupt and jerky, as in an old silent movie. She learns fast, however. In seven months, if all goes well, she will sit upright without support. A month later the infant can stand unassisted, but on average it takes seven more months for her to walk smoothly without consciously thinking of the action.

For the toddler to stand, the muscles that oppose each other in hips, knees, and ankles must exert an equal and opposite tension, stabilizing the joints and preventing them from folding up. "Muscle tone" describes the complex set of interactions that keeps all the infant's muscles in a mild state of contraction, making her erect posture as active and strenuous as the movements that follow it. The toddler's body crackles with millions of messages informing her brain of changing conditions and giving directions to perform the extraordinary feat of walking.

The brain stores our movements in a kind of hard disk in the lower brain. Repeat an action often enough and it becomes a subconscious reflex so that, for example, I can walk without thinking about it. This act of delegation explains Rubinstein's finger memory: the reflex response permits him to move his fingers faster than his higher brain could order them. It also explains the "waiter effect." Experienced waiters or flight attendants will not allow you to pick up a heavy coffee thermos from the tray, or the reflex would cause the hand holding the tray to jerk upwards, spilling the tray's other contents. On the other hand, they can lift the thermos without any danger of spilling, for the stored memory has learned to control the reflex. Predictability is the key. The same principle explains the tickling reflex: I cannot tickle myself because my own actions—though not someone else's—are predictable, drawing from this stored memory.

A MUSCLE CHOIR

Muscle cells perform just one action: they contract. They can only pull, not push, as their molecules slide together like the teeth of two facing combs. Cells join together in strands called fibers, and the fibers report to a further hierarchy called a motor unit group.

One motor nerve controls a motor unit group, wrapping itself around the muscle group as an octopus would encircle a pole. When that nerve gives a signal, all of its muscle fibers contract, becoming shorter and fatter. Muscles operate with an "all or none" principle. They have no throttle, rather a simple on-off switch. Strength varies, as when a pianist lightly taps a key or pounds it mightily, because of the number of motor units firing at any moment.

Conductors of large choirs warn their singers not to take breaths at the end of a pianissimo measure since the sound of many singers inhaling would be audibly distracting. Rather, they should suck in air in the middle of a measure, staggering their breathing so that the choir continues singing while just a few members inhale at any given instant. Similarly, the motor nerve directs some of a weightlifter's motor units to take a rest when needed, while the biceps' overall strength stays steady.

Rarely will all the motor units in a large muscle fire simultaneously. In an emergency, adrenaline may induce feats of great strength, called hysterical strength, such as a mother lifting a car off her child—perhaps then we galvanize all the motor units into simultaneous action.

I have literally heard the muscle "choir" by inserting a needle into a muscle and attaching it to a machine that transforms electrical energy into sound. Click-click-click: a constant stream of messages reports the activity of muscle tone. Slowly flex the biceps, and the volley of clicks accelerates. Move the arm abruptly, and the clicks crescendo to machine-gun frequency. The cells never stop clicking, and they adjust instantly, in a fraction of a second, when the brain calls for sudden movement.

As the meter records the stream of static flowing from just one muscle area the size of a needle point, hundreds of other muscles go undetected. A large and crucial group of them fire off whether or not we think about them: the automatic muscles controlling our eyelids, breathing, heartbeat, and digestion. The wise body does not trust our forgetful selves with these vital functions. So reliable are they that we cannot voluntarily stop our heartbeat or breathing. No one can commit suicide by holding their breath; accumulating carbon dioxide in the lungs will trigger a mechanism to override conscious desire and force the muscles of ribs, diaphragm, and lungs to move.

Consider the electrical network linking every home and building in a city like Beijing or New York. At any given second lights switch on and off, toasters pop up, microwave ovens begin their digital countdowns, water pumps lurch into motion. A far more complex switching system is operating in your body at this second, and it performs harmoniously, much of it at a subconscious level. When you reach the end of this page, you will turn or swipe it with your fingers, only vaguely aware of the elaborate systems that work together to permit such an act.

BALANCING ACT

In the physical body, as well as the spiritual, a muscle must be exercised. If a person loses movement through paralysis, atrophy will set in and muscles will shrink away, absorbed by the rest of the body. Similarly, the corporate Body of human beings shows its health best by acting in love and service to others. When it ignores cries of pain and injustice, and fails to respond, it begins to waste away.

As a Christian, I am struck by the disorderly way in which that particular Body moves. Study any century, and the history of the church then will include splits and divisions, heated debates about the role of social concern, and sadly excessive reactions. Because church history includes these spasms, we easily discount the effectiveness of the Body's motion.

As I look closer at the biology of motion, though, I can better grasp how seemingly disconnected spurts of energy can contribute to progress. In the human body every action has an equal and opposite reaction. Muscles pair up antagonistically so that when the triceps contracts the biceps relaxes, and vice versa. One of the pioneers of neurophysiology, Sir Charles Sherrington, demonstrated that *all* muscular activity involves inhibition as well as excitation. Every muscular sentence includes a clause with a balancing "but."

The knee-jerk reflex, which involves only two muscles, illustrates Sherrington's principle. When a doctor taps a patient's knee, muscles on the front of the thigh spring into action, excited. That response only occurs because muscles on the back of the thigh, which keep the knee bent, do not contract, an act of inhibition. The knee jerk relies on two equally powerful stimuli, one of which leads to action, the other to inaction. In complex movements, like walking or kicking a ball, hundreds of opposing reactions occur simultaneously. All muscular action follows this principle of give-and-take.

The biological process may help explain what at first glance appears as a troubling recurrence in the history of the church. The Body of Christ has sometimes moved by extreme, exaggerated reflexes. In behavior, as Charles Williams has pointed out, there are two opposite tendencies. "The first is to say: 'Everything matters infinitely.' The second is to say: 'No doubt that is true. But mere sanity demands that we should not treat everything as mattering all that much.'" The rigorous tendency leads to a sharp, intense view of the world that sees all actions as having eternal consequences. In its extreme forms it can evolve into pharisaical legalism and the fanaticism of "holy" crusades. The more moderate, relaxed approach can, at its worst, drift toward inactivity, a "who cares" attitude toward injustice and wrongdoing.

The apostle Paul, notably in Galatians and Romans, fought a pitched battle against both extremes, on the one hand denouncing legalists for perverting God's grace and on the other hand upholding Christian works as a normal outgrowth of new life.

Christians have vacillated between opposing forces. In the earliest Christian centuries, the Way of Affirmation and the Way of Negation sprang up, each attracting ardent followers. The negators retreated to the desert and demonstrated their spirituality in feats of self-denial, while the affirmers labeled those who abstained from marriage and feasting as "blasphemers against creation." The conflict was hardly new: Jesus contrasted John the Baptist's asceticism with his own reputation as a winebibber and glutton (Matthew 11:19). Each tendency produced something worthwhile: The Way of Affirmation gave us great art and romantic love and philosophy and social justice while the Way of Negation contributed the profound documents of mysticism that could only come from undisturbed contemplation.

If I visit a community of young radicals who advocate withdrawal from society and intentional poverty, I may come away with a distorted view of what compassionate activity in the world should look like. Are we not called to foster the common good, to look for ways to help others thrive? Yet such a counterculture may temper the activity of the institutional church, calling it back to a prophetic awareness of justice. Perhaps their contribution may keep the Body from toppling over to one side.

Richard Mouw of Fuller Seminary recalls being in a meeting with sociologist Peter Berger. Speaking as a seminary president should, Mouw said that every Christian is called to engage in radical obedience to God's program of justice, righteousness, and peace. Berger responded that Mouw was operating with a rather grandiose notion of radical obedience. Somewhere in a retirement home, he said, there is a Christian woman whose greatest fear in life is that she will make a fool of herself because she will not be able to control her bladder in the cafeteria line. For this woman, the greatest act of radical obedience is to place herself in the hands of a loving God every time she goes off to dinner.

On reflection, Mouw agreed that Berger had made a profound point. God calls us to deal with the challenges before us, and often our most radical challenges are very "little" ones. The call to radical micro-obedience may mean patiently listening to someone who is boring or irritating, or treating a fellow sinner with a charity that is not easy to muster, or offering detailed advice on a matter that seems trivial to everyone but the person asking for the advice.

C. S. Lewis was surprised to learn that his life after conversion consisted mostly in doing the same things he had done before, only in a new spirit. He concluded that being a practicing Christian "means that every single act and feeling, every experience, whether pleasant or unpleasant, must be referred to God." The unifying factor in Bodily motion must be a common commitment to the Head. We will disagree on interpretations of what Jesus said and what are the best means of accomplishing those goals. But if we fail to find fellowship in our mutual obedience to Christ, our actions will be seen not as the counterbalancing forces necessary for movement but as spastic, futile contractions.

THE ANARCHIC NECK

A blubbering hulk of a man entered my office in India. He was a successful Australian engineer who had been working in India, but now his neck twitched so violently that every few seconds his chin smashed into his right shoulder. He had spasmodic torticollis, or twisted neck syndrome, a debilitating condition sometimes caused by a psychological disorder.

Between the spastic flings of his chin, my patient described his despair. The torticollis, he said, had begun soon after a visit to Australia. A confirmed bachelor throughout his time in India, he had returned from Australia with a wife—a gorgeous woman, taller and younger than he, who soon became the subject of much village gossip. Short and obese, the engineer had a known history of alcoholism. What had she seen in him? What had prompted such a mismatch?

I referred the engineer to a psychiatrist, for I could do little except sedate him temporarily. The psychiatrist confided to me his suspicion that the engineer's condition had developed out of anxiety over his new wife. The patient returned to me in a few weeks, unkempt, his neck wrenching spasmodically, full of despair.

When he sat alone, unnoticed by anyone, his neck rarely contorted. As soon as someone struck up a conversation with him, however, his chin would slam into his shoulder, aggravating a chronic, spongy bruise. Nothing helped other than sedation and the temporary relief that followed an injection of his nerve roots with novocaine. Finally, he reached the point of utter despondency and attempted suicide. He insisted, with a firm and resolute edge to his voice, that he would try again and again until he succeeded. He could no longer continue living with his anarchic neck.

Since we had no neurosurgeon, I reluctantly agreed to attempt a dangerous and complicated operation that involved exposing his spinal cord and the base of his brain. I had never ventured a procedure quite so complex, but the man insisted he had no alternative other than suicide.

I cannot recall an operation plagued with as many mishaps. The cautery short-circuited at the critical time when we most needed it to control bleeding. Then all the hospital lights failed, and I was left with a hand-held flashlight and no cautery just when the spinal cord was coming into view. To add to the stress, I had neglected to empty my bladder and felt most uncomfortable throughout the surgery.

Amid these distractions, I tried to concentrate on some very delicate cutting. After exposing the spinal cord and lower brain, I traced the hair-like nerves that supplied the spastic muscles in his neck. Any slight quiver of the scalpel could have cut a bundle of nerves, destroying movement or sensation.

Somehow, in spite of these difficulties, the surgery proved successful. When the engineer awoke, his back humped with a bandage, he discovered that the feared neck movement no longer plagued him. It couldn't, of course, for I had cut the motor nerves that led from the spinal cord to the muscles that turned his neck. He could no longer make the movement that had previously dominated him.

SELF-SERVING MUSCLES

When people see someone with a spastic muscle, they often assume the muscle itself is malfunctioning. Actually, the muscle is perfectly healthy, and usually well-developed because of frequent use. The malfunction stems from the muscle's relationship to the rest of the body; it demonstrates its motility at the wrong times, when the body neither needs nor wants it. A spastic muscle may, as in the case of the Australian engineer, cause embarrassment, pain, and deep despair.

Quite simply, a spastic muscle disregards the needs of the rest of the body, its dysfunction more mutiny than disease. Sir Charles Sherrington studied a brainless frog swimming easily across a pond. You can, he said, get the impression the injury is trivial until you examine the behavior closely and see that the frog is swimming randomly, with no purpose, just kicking its legs as a reflex. Absent a brain, movement can have no "purpose."

Acts of love—healing, feeding, educating, ministering to prisoners, proclaiming the good news of God's love—are the spiritual Body's proper movements. Yet even these motions, which appear wholly good, can fall prey to a dangerous dysfunction. Like the spastic muscle, we may perform acts of kindness for the benefit of our own sakes and reputations. Those of us in Christian work face this constant tendency toward pride. Someone comes to me for spiritual counsel, and I give it. Before they have walked out of my room I'm congratulating myself on what a fine counselor I am.

Jesus' disciples, the first ones chosen to represent him, stumbled at this very point. They argued about petty issues: Who is the greatest disciple? Who will have the greatest honor in heaven? (Matthew 20:20-23). Jesus lectured them on the need for self-sacrifice, pulled children from the crowd to show the meek attitude they should have, and even washed the disciples' dirty feet to illustrate the ideal of service. It did not seem to sink in—not until after that dark day on Calvary.

I have no desire to judge Christians today who seem to be exercising their muscles in a self-serving rather than a Body-serving way. I do wonder, though, about the dangers facing megachurches in places like

Brazil and Africa, and especially the electronic church in the United States. A powerful "muscle" can reach millions of people and also attract millions of dollars in revenue. Does the medium give some leaders too much leverage and power? As a former missionary in a helping role, I know too well the human weaknesses that lead to spiritual pride. Those in the spotlight—media personalities and Christian speakers and performers—have described to me the besetting temptations that accompany ego strokes from adoring fans.

None of us is exempt. Radical Christians who urge social action, politically conservative Christians who give large sums of their investments to missions, seminary students who glory in their newfound knowledge, church members who join key committees—all of us need to come back to the image of God's Son kneeling on a hard floor and unbuckling sandals covered with grime and dust.

We are not called to display individual strength as a discrete unit in the Body. Rather, our activity must be for the sake of the whole Body. If in the process of serving, applause or even fame results, we will need special grace to handle it. And if we consciously seek renown or wealth, the effect will be like the spastic contraction of a once-healthy muscle. Like Ananias and Sapphira (Acts 5), we will have turned a good act into an impure act because of our impure motives.

Movement in the Body requires the smooth and willing cooperation of many parts who gladly submit their own strength to the will of the Head. Otherwise their actions, though powerful and impressive, will not benefit the whole.

What does this reliance consist of? Does the Spirit actually help with the specific pressures and choices confronting me each day? I explain my dilemmas and pour out my needs, but God does not respond by telling me what to do. There is no shortcut, no magic, only the possibility of a lifetime search for intimacy with a God who gently communicates to us through the Holy Spirit.

Paul urged fellow members, cells in Christ's Body, to learn God's "good, pleasing and perfect will" (Romans 12:2). In another passage he defined what our attitude should be, a reflection of Jesus' own:

In your relationships with one another, have the same mindset as Christ Jesus:

Who, being in very nature God,
  did not consider equality with God something to be used to his own
    advantage;
rather, he made himself nothing,
  by taking the very nature of a servant,
  being made in human likeness. (Philippians 2:5-7)

By following that model of humility I am, in fact, learning the mind of Christ, the Head of the Body.

PART FIVE

The LANGUAGE of PAIN

*The merest schoolgirl, when she falls in love, has
Shakespeare or Keats to speak her mind for her;
but let a sufferer try to describe a pain in his head
to a doctor and language at once runs dry.*

VIRGINIA WOOLF

CHAPTER SIXTEEN

$\mathscr{A}$ SENSE *of* PROTECTION

IN A LITHOGRAPH by nineteenth-century printmaker Honoré-Victorin Daumier, a distinguished gentleman in a white waistcoat is sitting on a high-backed Victorian sofa. Perhaps I should say contorting, not sitting, for pain has doubled him up. His legs bend under him and his back arches downward, nearly forming a fetal position. Four sets of leering little devils perch beside him, half of them playing tug-of-war with cables looped around the man's midsection and half sundering his abdomen with a huge, jagged-tooth saw. The man's face expresses absolute agony.

Daumier added a title to his drawing: *La colique,* the pain of colic. Like most viewers, I find it difficult to view the print without a wince, mirroring the poor man's anguish. Who has not felt at least a twinge from a muscular spasm caused by intestinal blockage or distension?

Pain is the hallmark of mortality. We plunge into the world through a woman's stretched and torn tissues, our first response a cry of fear or grief or both. Years later we exit the world, often in one last paroxysm of pain. Between those two events we live out our days, with pain always lurking at the door. The word itself derives from *poena,* the Latin word for punishment, a dark hint that the demons working the saw are more than imaginary.

Ironically, I have spent my medical career among people whose faces also bear the signs of punishment and anguish, but for the opposite reason. Leprosy patients suffer because they feel no pain. They long for the demons who would alert them to impending danger.

My fascination with pain began, I suppose, in my childhood. As we traveled the mountains of South India, my parents would bring along a few pairs of dental forceps. I interrupted my play to stand by, with wide eyes and a racing heart, as my mother or father extracted teeth—without anesthetic. I would watch my tiny mother wriggle her pointed forceps up between the gum and tooth, seeking a firm grip so the crown of the tooth would not break off when she yanked. She hung on fiercely to those forceps while the patient's own thrashing motions worked the tooth loose. The patients cried out, danced around wildly, and spit up blood. Still, even after seeing those reactions, onlookers lined up for treatment. Ridding themselves of toothache warranted the cost.

Occasionally in villages on the plains we would also see the impressive *fakirs*, religious men who demonstrated their conquest of pain. Some would push a thin, stiletto-type blade through their cheek, tongue, and out the other cheek, then withdraw the blade without bleeding. Others strung themselves high in the air by pulling on ropes that passed through a metal ring at the top of a pole and ended in meat-hooks stuck into the flesh of their backs. Showing no signs of pain, they dangled like spiders above an admiring crowd. Still others garishly decorated themselves with scores of oranges attached to large safety pins, which they jabbed into their skin. They laughed and merrily danced down the streets on stilts, jiggling the oranges in time with music.

On my return to India as an orthopedic surgeon, I encountered the full spectrum of human misery. Unaccountably, I found myself drawn to those who never came to the hospital, the deformed beggars who lined the entrances to temples, railway stations, and most public buildings. I saw clawed hands with missing fingers, ulcerated feet, paralyzed thumbs, and every conceivable kind of orthopedic defect, and I learned that no orthopedist had ever treated them or their fifteen million fellow sufferers worldwide. Because of the stigma attached to leprosy, few hospitals would admit them.

I have been studying pain ever since because leprosy destroys the nerves that carry pain, making the body devastatingly vulnerable to injury. I even came to appreciate those fanciful creatures who wielded the cables and saws in Daumier's lithograph. Are they truly demons? Without their apparent torments, would the gentleman attend to his colic?

THE BODY'S HOTLINE

As I write, nerve cells are informing my brain that my strained back needs attention. The nerve endings of pain receptor cells detect pressure or inflammation, translate that sensation into a chemical and electrical code, and send those messages to the brain, which in turn weighs their significance and dictates a response.

Pain messages travel along a hotline, insisting on priority. They can preoccupy the brain, in the process drowning out all pleasurable sensations. The entire body responds. The muscles in my aching back contract, which intensifies pain by squeezing nerves. My blood flow changes: blood pressure reacts to pain just as it reacts to anxiety and fear. I may go pale or flush or even faint. Pain may upset my digestion, causing a spasm that brings on the feeling of nausea.

At another level, pain may dominate me psychologically. I may become cranky, complaining to my colleagues and family. Perhaps I'll cancel an overseas trip to give my back extra rest, which in turn leads to further complications: guilt from letting people down or depression about my inability to work.

Amazingly, the sensation that evokes such a powerful response in every part of my body and mind soon fades into oblivion. Think back to your worst experience of pain and try to remember what it felt like. You cannot. You can summon up sharp recollections from other senses, such as the face of a childhood friend or the tune of the national anthem or even a memory of taste or smell piquant enough to prompt salivation. Yet the sense of tyrannical pain has somehow vanished. You have forgotten.

Dominating, subjective, and ephemeral, pain offers a research target as elusive as the quark. What is pain? When is it really there—and where?

As a medical student in England I had the rare privilege of studying under Sir Thomas Lewis, a pioneer in the study of pain. I remember those days well because Sir Thomas used his students as guinea pigs in his research. One gets an unforgettable perspective on pain by recording sensations while being pinched or pricked. Often Lewis subjected himself to the same tests, lest he misinterpret the students' accounts. He collected his findings into a book, *Pain,* that became a classic, a model of beautiful language as well as medical research.

We allowed blood pressure cuffs to be inflated around a metal grater that pressed into our arms, endured drippings of hot sealing wax, and dutifully performed isometric exercises while a tourniquet cut off our blood supply. One wicked contraption shot electrical voltage through the fillings of a tooth. Some volunteers submersed their hands in ice water, then hot water. We had cheeks and hands pricked simultaneously to determine which pain extinguishes the other. We heard bells rung and stories read aloud, and we repeated sequences of numbers in forward and reverse order, all to measure how distraction modifies the sensation of pain.

These exhaustive methods yielded a few basic measurements. At what point does it start hurting (pain threshold)? Do you ever grow accustomed to the heat or pressure (adaptation to pain)? Where does it hurt (distribution of pain)? At what point can't you endure it any more (pain tolerance)? Subjects also had to describe each pain verbally and distinguish degrees of pain (as many as twenty-one were reported).

We students came away with slight lesions, blisters, and pinpricks, as well as a diploma that exempted us from further victimization. The professors came away with graphs mapping out sensitivities on every square centimeter of the body. Such experiments have proceeded unabated for more than a century for one reason: the nervous system is incredibly complex. Each tiny swatch of the body has a different perception of pain.

I need not reproduce the charts here, for instinctively everyone knows the principles of pain distribution. A single speck of dust flies into your eye. You react immediately: your eye tears up, and you squint it shut and dab at the eyelid to remove the speck. Such a speck can immobilize even a superbly conditioned athlete like a baseball pitcher; the pain is so great that he cannot continue pitching until the speck is removed. The same speck on the pitcher's arm would go wholly unnoticed. Indeed, thousands of dirt particles will accumulate there in the course of the game. Why the disparity in sensitivity?

The eye has certain rigid requirements of structure. Unlike the ear, its well-guarded sensory neighbor, the eye must lie exposed on the surface, in a direct line with light waves. The eye must also be transparent, which severely limits a blood supply, for opaque blood vessels would block vision. Any intrusion represents danger since the blood-starved eye cannot easily repair itself. Therefore a well-designed pain system makes the eye extraordinarily sensitive to the slightest pressure or pain, and its hair-trigger response trips the blink reflex.

Every bodily part has a unique sensitivity to both pain and pressure, determined by function. The face, especially in the area of the lips and nose, is acutely sensitive to both. Feet, subject to a day's stomping, are better protected by tough skin and thus mercifully insensitive. Fingertips present an unusual case. Constant use requires them to be sensitive to pressure and temperature but relatively pain resistant: carpenters would be rare if the gripping fingers fired off pain signals to the brain at every stroke of a hammer. In the body's torso, protecting vital organs becomes the main concern. Thus a light tap on the foot goes unnoticed, on the groin is felt as painful, and on the eye causes agony.

As I study pain in the human body, I marvel again at the Creator's wisdom. I might prefer that the lining of the trachea were even more sensitive to irritants, causing more pain and coughing and so making lung-destroying tobacco smoke intolerable. But could humans even endure a hypersensitive trachea in a dust storm or in our modern polluted environment?

I think again of the eye and its split-second response. Wearers of contact lenses might wish for less sensitivity in the eye, but the sensitivity benefits the great majority of people and their need to preserve vision. Each part of the body responds to the appropriate danger that might interfere with it and thus affect the whole body.

When I began to see patients, I encountered the phenomenon of referred pain. The economical body assigns pain sensors to alert us to the most common threats. The intestine warns of distention, though not cutting or burning; the secluded brain has few pain receptors. If a part of the body faces an unexpected danger, the body borrows pain sensations from other regions. An injured spleen may seek help from faraway pain receptors in the tip of the left shoulder, and a kidney stone may be felt anywhere along a band from the groin to the lower back.

Referred pain makes proper diagnosis of a heart attack a tricky problem for a young doctor. "It's a burning sensation in my neck," one patient reports. "No, it feels like my arm is being squeezed," says another. In a sense, the spinal cord is playing a trick on the brain. A warning system located in the spinal cord or lower brain detects a cardiac problem and instructs unrelated skin and muscle cells to act as if they are in serious jeopardy, as a favor to their wordless neighbor. Remarkably, the borrowed area—say, the left arm—may even feel tender to the touch. The left arm puts on an acting performance in order to seize the attention of a victim who would otherwise not attend to a vulnerable heart.

I learned to see pain as a kind of language, the most effective language in mobilizing the body's response to potential harm.

ONGOING CONVERSATION

Daily, pain contributes to our quality of life, even in such a common activity as walking. A leprosy patient, with perfectly normal skin tissue on the soles of his feet, may return from a walk with foot ulcers. A healthy person who takes exactly the same walk will develop no blisters or ulcers. Why? A file cabinet in my office contains a box of photographic slides that illustrate the reason. The stress of walking causes an increase in blood supply and mild inflammation, which we can measure through a thermograph machine that displays heat in various colors.

The slides of color-coded feet show that the way a healthy person puts feet to the ground changes radically from the first mile to the fifth mile. Perhaps at the beginning your great toe absorbs most of the stress; by the end of the walk your lateral toes and the lateral border of your foot will take over. Later the toe and heel will come down together. When you begin a really long hike, you will start off heel-toe, heel-toe. But when you return, you'll be lifting your foot and setting it down as one unit—all these adjustments having been made subconsciously.

You do not make those shifts because of muscular fatigue. Rather, pain cells in your toes, heels, arches, and lateral bones have intermittently informed the brain, "Ease up a little. I need some rest." You stride along oblivious since your brain assigns these functions to a subliminal control system that constantly monitors pain and pressure in every part of your body. A leprosy patient, having lost this incessant hum of intercellular conversation, will walk five miles without changing gait or shifting weight. The same pressure strikes the same cells with unrelenting force. In the slides of my patients, the favored portions of the foot show a white-hot color, a visual warning of potential ulcers.

As I sit and write these words, the pain cells in my hips and legs are periodically asking me to shift my weight a bit, and I reflexively obey by changing position and crossing and uncrossing my legs. Pain employs a tonal range of conversation. It whispers to us in the early stages of risk; subconsciously, we feel a slight discomfort and toss and turn in bed. (A paraplegic person lives in constant fear of bedsores because he or she can no longer hear those whispers.) It speaks to us as hazard increases: a hand grows tender and sore after a long stint at raking leaves. And pain shouts at us when the danger becomes severe: blisters, ulcers, and tissue damage force us to pay attention.

A limp amplifies the body's response to pain. Out of orthopedic habit, I tend to stare impolitely at people who limp. What they may view as an embarrassing malfunction, I view as a wonderful adaptation. A limper's body is compensating for damage to one leg by redirecting weight and pressure to the other, healthy leg. Every normal person limps occasionally. Sadly, leprosy patients do not limp, and their injured legs never get the rest needed for healing.

When the body's pain-monitoring system breaks down, the inability to sense pain can cause permanent damage. Perhaps you step on a loose stone or curb. As your ankle begins to twist, the lateral ligaments of the ankle endure a terrific strain. Detecting the strain, nerve cells order the body to take all weight off the damaged leg, and its thigh and calf muscles will become momentarily flaccid. If your other, undamaged leg has lifted off the ground to take a step, you will now have no support and will lurch to the ground. (A step, says the anatomist, is a stumble caught in time.) Your body opts for falling rather than forcing the ankle to take on weight in its twisted position. You get up feeling a fool and hoping no one was watching, but in reality you have just achieved a beautifully coordinated maneuver that saved you from a badly sprained ankle or worse.

I cannot erase from my mind the memory of watching a leprosy patient sprain his ankle without falling. He stepped on a loose stone, turned his ankle *completely* over so that the sole of his foot pointed inward, and walked on without a limp. Lacking the protection of pain, he did not even glance at the foot he had just irreparably damaged by rupturing the left lateral ligament! Despite his therapist's warnings, in subsequent days he kept turning his ankle again and again until eventually, due to more complications, he had to have that leg amputated.

Pain, so often viewed as an enemy, is actually the sensation most dedicated to keeping us healthy. If I had the power to choose one gift for my leprosy patients, I would choose the gift of pain.

CHAPTER SEVENTEEN

The UNIFIER

WHEN I REFLECT ON PAIN, I prefer not to think in a detached, aca-
demic way. Instead, I focus on one individual, and at such moments my
mind flashes back to the refined features of Sadagopan, whom we called
Sadan. He personified the soft-spoken, gentle Indian spirit.

When Sadan arrived at Vellore, his feet had worn away from injury to
half their normal length and his fingers were shortened and paralyzed. It
took two years of tireless effort to halt the pattern of destruction in his feet.
Meanwhile, we began reconstructing his hands, one finger at a time, attaching the most useful tendons to the most useful digits and retraining
his mind to adjust to the rewired connections. In all, Sadan spent four years
in rehabilitation from the many surgeries. Together, we wept at our failures
and rejoiced at the gradual successes. I came to love Sadan as a dear friend.

At last Sadan decided he should return home to his family in Madras
for a trial weekend. He had come to us badly marred by the disease,
which made him an outcast. Now his hands were more flexible, and with
a customized rocker-type shoe he could walk without damage. "I want
to go back to where I was rejected before," he said proudly, referring to
the cafés that had turned him away and the buses that had denied him
service. "Now that I am not so deformed I want to try my way in the great
city of Madras."

Before Sadan left, we reviewed together the hazards he might encounter. Since he had no warning system of pain, any sharp or hot object

could harm him. Having learned to care for himself in our hospital and workshop, he felt confident as he boarded the train to Madras.

That first night, after an exuberant reunion dinner with his family, Sadan retired to his old bedroom, where he had not slept for four years. He lay down on the woven pallet on the floor and drifted off to sleep, peaceful and contented. At last he was home, fully accepted once again.

The next morning when Sadan awoke and examined himself, as we had trained him to do, he recoiled in horror. The back of his left index finger was mangled. Instantly he knew the reason because he had seen such injuries on other patients. The evidence was clear: prints in the dust, telltale drops of blood, and of course, the decimated clump of tendon and flesh that had been so carefully reconstructed some months before. A rat had visited him during the night and gnawed his finger. Made insensitive by leprosy, he had felt nothing.

What will Dr. Brand say? he thought. All that day he agonized. Should he return to Vellore early? He decided he must keep his promise to stay the weekend. He looked for a rat trap to protect him that last night at home, only to find the shops closed for a festival. *I must stay awake all night*, he decided, in order to guard against further injury.

The next night Sadan sat cross-legged on his pallet, his back against the wall, studying an accounting book by the light of a kerosene lantern. Around four o'clock in the morning the subject grew dull, his eyes felt heavy, and he could no longer fight off sleep. The book fell forward onto his knees, and his hand slid over to one side against the hot glass of the hurricane lamp.

When Sadan awoke the next morning, he saw to his dismay that a large patch of skin had burned off the back of his right hand. He sat up, trembling, overcome with despair, and stared at his two hands—one gnawed by a rat, the other melted down to the tendons. He knew well the dangers and difficulties of leprosy, and in fact had taught them to others. Now he was devastated by the sight of his two damaged hands. Again he thought, *How can I face the doctors and therapists who worked so hard on these hands?*

Sadan returned to Vellore that day with both hands swathed in bandages. When he met me and I began to unroll the bandages, we both wept. As he poured out his misery to me, he said, "I feel as if I've lost all my freedom." And then, a question that has stayed with me, "How can I ever be free without pain?"

SELF-UNITY

Sadan's plight is shared by millions of people who suffer from leprosy and other numbing diseases: diabetes, for instance, can have a similar effect on the extremities. Insensitivity offers a stark lesson about the value of pain. At its most basic level pain serves as a warning signal, like a smoke alarm that goes off with a loud noise when it senses fire. Sadan nearly lost his hands because he lacked that signal.

Pain makes another related contribution that often gets overlooked: it unifies the body. In truth, Sadan suffered because the rest of his body had lost contact with his hands. A body possesses unity to the degree that it feels pain. An infected toenail reminds me that the toe belongs to me, an integral part of my body's health. Hair—yes, that matters, but mainly as a decoration. We can bleach, shape, iron, and even shave it off without pain. What is indispensably *me,* pain defines.

Nothing distresses me more than watching my patients in the Louisiana hospital lose contact with their own hands and feet. As pain fades away they begin viewing their own limbs as stuck-on appendages. You and I speak metaphorically of a hand or foot going "dead" when we sleep on it in an awkward position. Leprosy patients seem to regard their hands and feet as truly dead.

The most common injury at Carville, the "kissing wound," occurs when a cigarette burns unnoticed down to the nub and brands matching scars into the skin between the two fingers. Leprosy patients think of their hands as accessories, not unlike a plastic cigarette holder. One such patient, who was carelessly destroying his hands, said to me, "You know, my hands are not really hands—they're things, just like wooden attachments. And I always have the feeling they can be replaced because they are not me."

I have to keep reminding leprosy patients to attend to painless parts of their bodies they may otherwise disregard. Although I try desperately to awaken in them a sense of their bodies' unity, overcoming their detachment seems impossible apart from pain. Just as pain unifies the body, its loss irreversibly destroys that unity.

In India I had one group of teenage patients nicknamed "the naughty boys" because they tested the limits of our medical longsuffering. These rascals competed to shock others with their displays of painlessness. They would thrust a thorn all the way through a finger or palm, pulling it out the other side like a sewing needle. They juggled hot coals or passed their hands over a flame. When quizzed about a wound on hands or feet, they grinned mischievously and said, "Oh, it must have come by itself."

Eventually, after taxing all our skill in education and motivational therapy, most of the "naughty boys" gained a respect for their bodies and learned to redirect their ingenuity to the goal of preserving their hands and feet. Throughout the rehabilitation process I felt as if I was introducing the boys to their limbs, urging them to reclaim these numbed parts.

Years later when I began working with laboratory animals, I learned that they had even more estrangement from deadened parts of their bodies. If I denervated rats for an experiment, I had to keep them well fed; otherwise, the next morning I would find these animals with shortened feet and legs. I am told that a wolf or coyote, losing sensation through frostbite or a trap injury, will gnaw through its leg and limp away unperturbed. That image captures for me the worst curse of painlessness: the painless person, or animal, loses a fundamental sense of self-unity.

ROOTS OF COMPASSION

An amoeba, one-celled, perceives any threat as a danger to the whole. Bodies consisting of many cells need something more, and pain provides that crucial unifying link. Individual cells must suffer with one another for multicellular organisms to survive; the head must feel the needs of the tail. In the human nervous system, one slender nerve cell connecting the toe to the spinal column may span four feet—no other cell in the human body approaches that length.

As I turn from the network of pain in biology to its analogy in a spiritual Body, again I am struck by the importance of such a communication system. Pain serves the same vital role in uniting a corporate membership as it does in guarding the cells of my own body. A healthy body feels the pain of its weakest part.

Naturally, there are differences between the unity attainable in a physical body of linked cells and in a Body composed of autonomous members. No tangible axons stretch from person to person in, say, the global church. Nevertheless, a healthy spiritual Body shares the pain of all its members. When wounded, living tissues cry out, and the whole body hears the cry. And we in Christ's Body—loving our neighbors as ourselves—are called to a similar level of identification: "If one part suffers, every part suffers with it," says the apostle Paul (1 Corinthians 12:26).

Deep emotional connections link human beings as surely as neurons (nerve cells) link the parts of our bodies, as shown even in sporting events. Watch the face of a wife sitting in the stands at Wimbledon as her husband plays in the championship tennis match. The action on the court can be read on the wife's face. She winces at every missed shot and smiles at each minor triumph. What affects him affects her.

Or recall the effect on a nation when a beloved leader dies. I experienced the unifying effect of pain most profoundly in 1963 when I came to the United States to address the student chapel at Stanford University. As it happened, the chapel service occurred just two days after the assassination of President John F. Kennedy. I spoke on pain that day, for I could read nothing but pain on the faces of hundreds of students jammed into that building. I described for them scenes from around the world, where I knew groups of people would be gathering together in prayer and mourning to share the pain of a grieving nation. I have never felt such unity of spirit in a worship service.

Something like those sympathetic connections joins us to members of Christ's Body all over the globe. When an oppressive government jails courageous Christians, when Islamic radicals behead those of a different religion, or even when my neighbors lose their jobs, a part of my Body suffers and I sense the loss. The pain of others can also come to our attention in less dramatic ways: the whispered signals of loneliness, depression, opioid addiction, discrimination, physical suffering, self-hatred.

"How can a man who is warm understand one who is cold?" asked Alexander Solzhenitsyn as he tried to fathom the apathy toward millions of Gulag inmates. In response, he devoted his life to perform the work of a nerve cell, alerting us to pain we may have overlooked. In a Body composed of millions of cells, the comfortable ones must consciously attend to the messages of pain, cultivating a lower threshold of pain. The word *compassion* itself comes from Latin words *cum* and *pati*, together meaning "to suffer with."

Today our world has shrunk, and as a Body we live in awareness of many cells. A litany of suffering fills our media outlets. Do I fully attend? Do I hear their cries as unmistakably as my brain hears the complaints of a strained back or broken arm? Or do I instead filter out the annoying sounds of distress?

Closer, within my own local branch of Christ's Body—how do I respond? Tragically, those who are struggling with divorce, alcoholism, gender or sexual identity, introversion, rebelliousness, unemployment, or marginalization often report that the church is the last group to show them compassion. Like a person who takes aspirin at the first sign of a headache, we want to silence them, without addressing the underlying causes.

Someone once asked John Wesley's mother, "Which one of your eleven children do you love the most?" She gave a wise answer to match the folly of the question: "I love the one who's sick until she's well, and the one who's away until he comes home." That, I believe, is God's attitude toward our suffering planet. Jesus always stood on the side of those who were suffering; he came for the sick and not the well, the sinners and not the righteous.

God gave this succinct summary of the life of King Josiah: "He defended the cause of the poor and needy, and so all went well." And then this poignant postscript: "Is that not what it means to know me?" (Jeremiah 22:16).

ANOTHER KIND OF UNITY

Skeptics see divisiveness as the church's greatest failure. In response, leaders exhort denominations and interfaith groups to join hands in a national or worldwide campaign. From my experience with the body's nervous system, I would propose another kind of unity, one based on pain.

I can read the health of a physical body by how well it listens to pain—after all, most of the diagnostic tools we use (fever, pulse, blood cell count) measure the body's healing response. Analogously, the spiritual Body's health depends on whether the strong parts attend to the weak.

I have performed many amputations, most of them the result of a hand or foot no longer reporting pain. There are members of the corporate Body, too, whose pain we never sense, for we have denervated or cut whatever link would carry an awareness of them to us. They suffer silently, unnoticed by the rest of the Body.

I think of my Palestinian friends, for example. In places like Bethlehem, children have grown up knowing nothing but occupation and war. They play not in parks but in crumbling buildings pockmarked by rifle fire and explosives. Palestinian Christians feel abandoned by the church in the West, which focuses so much attention on Israel and assumes all non-Israelis in the Middle East to be Arab and Muslim. Beleaguered Christians in places like Lebanon and Syria plead for understanding from their brothers and sisters in the West, but we act as though the neuronal connections have been cut, the synapses blocked. Few hear their pain and respond with Christian love and compassion.

I think of the LGBTQ population scattered throughout our churches and colleges. Surveys show that a significant percentage of students in Christian colleges struggle with same-sex attraction. Yet some college administrations simply pretend those issues do not exist. Those students are left to flounder, cut off from the balance and diversity of the larger Body and from the acceptance and understanding that they need.

I think of the elderly, often put away out of sight behind institutional walls that muffle the sounds of loneliness. Or of battered children who grow up troubled, unwelcomed into foster homes. Or of refugees who feel cut off from participation in the larger Body. Or of prisoners sealed off behind electric fences. Or of foreign students who live in enclaves of cheap lodging, or of the homeless who lack any lodging at all.

Modern society tends to isolate these problems by appointing professional social workers to deal with them. No matter how well-intentioned, institutionalized charity can isolate hurting members from close personal contact with healthy ones. Both groups lose out: the charity recipients who are cut off from person-to-person compassion and the charity donors who offer care solely as a material transaction.

In the human body, when an area loses sensory contact with the rest of the body, even when its nourishment system remains intact, that part begins to wither and atrophy. In the vast majority of cases—ninety-five of every hundred insensitive hands I have examined—permanent injury or deformation results. Likewise, loss of feeling in the spiritual Body leads to atrophy and deterioration. So much of the sorrow in the world is due to members who simply do not care when another part suffers.

The apostle Paul set out clearly how members of Christ's Body should respond to those who suffer:

> Praise be to the God and Father of our Lord Jesus Christ, the Father of compassion and the God of all comfort, who comforts us in all our troubles, so that we can comfort those in any trouble with the comfort we ourselves receive from God. For just as we share abundantly in the sufferings of Christ, so also our comfort abounds through Christ. . . . And our hope for you is firm, because we know that just as you share in our sufferings, so also you share in our comfort. (2 Corinthians 1:3-5, 7)

Many accounts of Christians who have suffered, beginning with the book of Job and the Psalms and continuing through the writings of the saints, speak of a "dark night of the soul" when God seems strangely absent. When we need God most, God seems most inaccessible. At this moment of apparent abandonment, the Body can rise to perhaps its highest calling: we become in fact Christ's Body, the enfleshment of his reality in the world.

When God seems unreal, we can demonstrate that reality to others by expressing Christ's love and character. Some may see this as God's failure to respond to our deepest needs: "My God, why have you forsaken me?" I see it as a calling for the rest of the Body to unify and to embody the love of God. I say this carefully: we can show love when God seems not to.

PEDRO'S PALM

One of my favorite patients at Carville, a man named Pedro, taught me about developing greater sensitivity to pain. For fifteen years he had lived without the sensation of pain in his left hand, yet somehow the hand had suffered no damage. Of all the patients we monitored, only Pedro showed no signs of scarring or loss of fingertips.

While mapping sensation on Pedro's hand, one of my associates made a surprising discovery. One tiny spot on the edge of his palm still had normal sensitivity so that he could feel the lightest touch of a pin, even a stiff hair. Elsewhere, the hand felt nothing. A thermograph showed that the sensitive spot was at least six degrees hotter than the rest of Pedro's hand (bolstering our theory, still being formulated, that warm areas of the body resist nerve damage from leprosy).

Pedro's hand became for us an object of great curiosity, and he graciously obliged without protest as we conducted tests and observed his activities. We noticed that he approached things with the edge of his palm, much as a dog approaches an object with a searching nose. For instance, he picked up a cup of coffee only after testing its temperature with his sensitive spot.

Pedro eventually tired of our fascination with his hand. To satisfy our curiosity, he told us, "You know, I was born with a birthmark on my hand. The doctors said it was something called a hemangioma and froze it with dry ice. But they never fully got rid of it because I can still feel it pulsing."

Somewhat embarrassed that we had not considered that possibility, we verified that the blood vessels in his hand were indeed abnormal. A tangle of tiny arteries supplied an extra amount of blood and directed some of it straight back to the veins without sending it through all the fine capillaries. As a result, the blood flowed very swiftly through that part of his hand, keeping its temperature close to that of his heart, too warm for the leprosy bacilli to flourish.

A single warm spot, the size of a nickel, which Pedro had previously viewed as a defect, gave him a great advantage after he contracted leprosy. That one remaining patch of sensitivity protected his entire hand.

In a church that has grown large and institutional, I pray for similar small patches of sensitivity. I look for modern prophets who, whether in speech, sermon, or art form, will call attention to the needy by eloquently voicing their pain. A healthy Body values these pain-sensitive members, just as Pedro valued his tiny spot of sensitivity.

"Since my people are crushed, I am crushed," cried Jeremiah (Jeremiah 8:21). And elsewhere,

> Oh my anguish, my anguish!
> > I writhe in pain.
> Oh, the agony of my heart!
> > My heart pounds within me,
> > I cannot keep silent. (Jeremiah 4:19)

Micah, too, wrote of his grief at his nation's condition:

> Because of this I will weep and wail;
> > I will go about barefoot and naked.
> I will howl like a jackal
> > and moan like an owl. (Micah 1:8)

These prophets stand in great contrast to insensitive Jonah, who cared more about his comfort than about an entire city's destruction. The prophets of Israel sought to warn an entire nation of its social and spiritual numbness. I sense a need for modern Jeremiahs and Micahs—too often dismissed as young radicals—who can direct our attention to those in need of compassion and comfort.

By ignoring pain, we risk forfeiting the wonderful benefits of belonging to the Body. For a living organism is only as strong as its weakest part.

CHAPTER EIGHTEEN

CHRONIC PAIN

ON SEVERAL OCCASIONS I have met a pain that defies understanding. Although it serves no apparent purpose, it so tyrannizes a life that the patient can think of little else.

Rajamma crept into my office with the demeanor of a hunted animal. As if scouting for an enemy, she peered warily around the room before lowering herself into a chair. She had many enemies: a sudden noise, anything that might startle her, even a gust of wind. Her cheeks were sunken, and she was thin to the point of emaciation. Marks scarred her face in a peculiar pattern, which I recognized as treatments by the traditional medicine man. She had scratched and burned her facial skin so that it had taken on a tough, leathery texture.

At this point I had been working in India only a year or two. During my training in London, whenever I encountered a case requiring a specialist I promptly referred the patient to someone more qualified and experienced. In South India I had no such luxury.

Rajamma suffered from *tic douloureux,* neuralgia of the face in its most severe form. The pain attacks spasmodically, an overwhelming shot of agony to one side of the face. It causes a grimace, thus the name "tic," suggesting a twitch of the facial muscles. The condition may develop after an infection, such as a septic molar tooth, or it may have no identifiable cause. Even though Rajamma could not identify any tooth problems, local dentists had extracted all the teeth on one side of her face in hopes of eliminating the source of pain. As she told me her story, speaking slowly, she kept her mouth open and moved her lips carefully to limit the movement of her cheeks.

Her husband explained that the family tiptoed around and would not laugh or even tell a joke for fear of triggering one of her attacks. I asked about her near-starvation weight. "I dare not chew," she said, "so I live on fluids, not too hot, not too cold."

Rajamma lived in a tiny, earthen hut with her husband and four children. "My children never play in or near the house anymore," she said with regret. Though chickens usually have the run of village houses, she kept hers penned so that none would fly up or startle her by cackling.

Despite all these precautions Rajamma lived at the mercy of excruciating pain, which hit her many times a day, incapacitating her. In desperation she had the village "doctors" heat a metal tube in the fire and burn blisters onto her face in an attempt to quell the pain. Her mental health had deteriorated badly and, though her husband tried to sympathize, I could tell the family was approaching a state of crisis.

I made every effort to locate a physical cause for the pain, and failed. Twice I tried to deaden the trigger area, which seemed to lie just in front of her right cheek bone. The first time, the sight of the needle near her face touched off one of her most savage attacks. My second attempt, under anesthesia, did not succeed.

Reluctantly, I concluded that I had only one sure way to stop Rajamma's agony: I must open her skull and divide the nerves supplying that section of her face. I put off this decision because I had not been trained for neurosurgery and indeed had never observed such a procedure. But I saw no other recourse. Fortunately, in an anatomy course in Wales years before, I had dissected the cranial nerves and knew just where to find the Gasserian ganglion within the bony coverings of the brain. A nerve block to that ganglion represented her only hope.

I explained the procedure to Rajamma and her husband, emphasizing the dangers. I might fail because of my inexperience. Perhaps worse, I might cut more of the nerves than should be cut. In that event her eyeball, as well as her cheek, might become insensitive, which could eventuate in blindness. I painted a bleak picture of possible consequences, yet nothing I said to the couple caused the slightest flicker of hesitation. The impact of Rajamma's suffering on her family was so great that even if I told them she would lose an eye in the operation, they would have readily consented.

Over the next week I studied all the books I could find and planned a strategy with our anesthesiologist. Since I wanted to communicate with the patient during surgery, we chose an anesthetic that would keep her alert enough to respond to questions. The day for surgery arrived.

THE FAMILY AWAITS

We arranged Rajamma in a sitting position in order to minimize pressure on the veins in her head, and after the anesthetic had taken hold I began to cut. The Gasserian ganglion lies at the junction of the fifth cranial nerve, in a venous sinus surrounded by bone. Inside this cavity, veins and nerves crisscross in a skein of tangled threads, making it impossible to keep the site free of blood. I chipped away the overlying bone and entered the cavity, picking through the layers of tissue one by one. At last I could see the base of the cavity. A plexus of nerve tissue, an inch across and half an inch deep, lay glimmering under my light like a crescent moon. From that mass, fine white nerve fibers fanned out, like tributaries of a river, toward the face.

In most of the body, a tough sheath enwraps each nerve, allowing it to tolerate a certain amount of stretching. Nerves in the bony skull, not designed to be touched or pulled, lack this protective sheath, and the slightest tremor of my hand would tear a nerve irreparably.

I took special care to identify a motor nerve, any damage to which would partially paralyze her jaw. But the other fibers all looked the same and lay bundled so close together that I could not be sure which was which. I electrically stimulated one delicate fiber and asked Rajamma what she felt. "You are touching my eye," she said. My heart beat faster, and beads of sweat popped out on my forehead as I dropped that slender nerve back into place.

I stared into the spreading pool of blood, pale and watery from the anemia caused by Rajamma's malnutrition. (At that time we had no blood bank to enrich her blood before surgery.) Finally, I separated two tiny, white nerve fibers and lifted them away from the blood. These two seemed the most likely carriers of the pain impulses that were making her life a misery. The time had come for me to cut them.

As I lifted the two nerves with my probe, an awful sensation broke over me like a wave. I was transfixed by the import of what I was about to attempt. Surgeons are trained to maintain a certain distance from patients so that personal feelings will not impair our judgment—for this reason we are warned not to operate on our own family members. At that moment I had a vision of Rajamma's family gathered around me in a circle, staring, waiting to see what I would do with her life.

Was one of these two the faulty nerve? So little is known about the physiology of nerves that I could not possibly spot a visual defect. As I gazed at those trembling strands of soft white matter, each the thickness of a cotton sewing thread, I found it hard to believe they held such power. Yet these nerves, containing hundreds of axons serving thousands of nerve endings, were domineering an entire family. Meanwhile, nerves just like them were steadying my hands and letting me know exactly how much force to apply to my instruments.

With a start, I came to myself. Though my reverie lasted only five or ten seconds, I have never forgotten the vision brought on by minuscule, shimmering nerves. Not knowing which of the two carried the pain, I had to sacrifice both. I cut them with two snips. We quickly got the bleeding under control and closed the wound.

Back in the ward, after Rajamma awoke fully, we mapped out the area on her cheek that no longer had sensation. My knots of tension relaxed when we learned the insensitivity did not include her eye. Haltingly, Rajamma tried what previously had triggered her spasms of pain. She attempted a slight smile, her first intentional smile in years. Her husband beamed back at her. With a quizzical look, she scratched her right cheek, aware that she would never feel anything there again.

Little by little after that, Rajamma's world fell into place. She became a gentle, sweet person once again. Her husband's anxiety began to lessen. Back home, she welcomed chickens into the house. The children began to play indoors, to jump and chase each other even in their mother's presence. In ever-widening circles, life returned to normal for that family.

ROGUE PAIN

In my career I have encountered a mere handful of patients who, like Rajamma, suffer from intractable pain with no apparent physical cause. Only a few times have I had to silence pain surgically by cutting a nerve. Those of us in medicine view such a procedure as a radical one of last resort. It carries with it grave risks: the potential of denervating the wrong areas, danger to the body parts made insensitive, and most mysteriously, the chance that even after nerves are cut the pain may persist as "phantom pain."

In Rajamma's case I had to counter all medical instinct by treating pain itself as the problem, not as a valuable symptom. That change in perspective highlights the cruel paradox of chronic pain: no longer a directional signal that points to something else, pain becomes a self-perpetuating malevolence. Those who suffer from chronic pain care only about how to switch it off.

Most commonly, chronic pain occurs in the back, neck, or joints, although those who suffer from cancer and a few other diseases can experience such pain anywhere. Whereas painless people—such as my leprosy patients—yearn for the warning signal of pain, chronic pain sufferers hear a blaring, pointless alarm.

A flurry of recent research has focused on chronic pain, and hundreds of pain clinics now specialize in it. The preferred methods of treatment are moving away from older surgical techniques, and a rise in opioid addiction has dampened enthusiasm for chemical solutions. Instead, the term *pain management* has entered the vocabulary of specialists. The director of one of America's largest chronic pain clinics has said we may need to apply a different model to chronic pain. Perhaps, he suggests, we should treat chronic pain as we treat diseases such as diabetes, by teaching patients to live comfortably *despite* the disease.

The broader health care industry now offers up alternative treatments for chronic pain: foot or earlobe massage, bee-stinger acupuncture, numbing pads, biofeedback, self-hypnosis. Devices such as TNS (transcu-taneous nerve stimulators) offer a more technological approach. Most of these techniques of pain management rely on overloading the brain circuits with diversionary stimuli, which in turn suppress incoming pain signals.

I prefer simpler methods to accomplish the same purpose. For example, I recommend a stiff-bristle hair brush for a person experiencing arm or leg pain. The effect of briskly stroking the skin will excite touch and pressure sensors and often relieve pain. Or when my chronic back pain intensifies I go for a barefoot walk on the rough shell-and-gravel sidewalks near my home.

COMPASSION FATIGUE

Images of suffering fill our television screens daily, a form of chronic pain on a global scale. Jesus himself acknowledged the deep-rooted nature of human misery when he observed (in a statement that is often grossly misapplied), "The poor you will always have with you" (Mark 14:7).

Having lived in a country where suffering was a distressing reality, I know well the dilemma posed by chronic pain on a massive scale. I have stared at long rows of patients, knowing I must decline treatment for all but a handful, and knowing too that thousands more await attention in remote areas.

We tend to view global suffering indirectly, via news reports and magazine articles, and thus pain forces a choice upon us. We can choose to extend our aid and food and abundance to help ease human misery, or we can numb the chronic pain by averting our gaze from the problems. The Bible makes clear that we in the Body have a responsibility to the suffering of those in the margins. Overseas relief aid administered by Christian agencies has mushroomed in recent years, a sign that we are attending to the short-term, crisis pains of the world. Christians have helped spearhead emergency responses to crises in Asia, Central America, and Africa, contributing billions of dollars to support such efforts. We who are strong help the weak.

Nonetheless, in handling chronic, long-term pain, the church still seems in its infancy. The head of one large Christian relief agency confessed,

I must restrain myself from global ambulance-chasing. When a major disaster occurs that captures media attention, our donors respond with incredible generosity. Agencies like mine collect millions of dollars, and move in with a kind of overkill. When the crisis is "hot news," we have no difficulty raising funds. Six months later, the desperate problems are still there, but the camera crews have gone elsewhere, and no one cares about the long-term suffering.

Although intense suffering may prompt a sudden outpouring of aid, donors soon tire of hearing about depressing conditions. Instead of increasing sensitivity, as a human body does in response to injury, we decrease it. Our focus turns from "How do I deal with the cause of the pain?" to "How can I silence it?" No longer a stimulus for action, the pain becomes a dull, ineffective throb. It has worn us down.

The field of health services illustrates the dilemma of relief work. People readily donate money for hospitals, drugs, and medical supplies. Yet, according to the World Health Organization, the great majority of health problems—as much as 80 percent of all diseases—derive from polluted water supplies. Development programs for sanitation and hygiene simply don't have the drawing power of more dramatic appeals.

Of course, chronic pain also occurs close to home, not just in places like Ethiopia and the Sahel. During tough economic times the United States and Europe also hear plaintive cries from people who cannot provide for their own basic necessities. That sound, too, can become a dull throb, easier to tune out than attend to.

A few years ago, as an economic slump and the impact of budget cuts in social programs began to affect people in urban areas, churches found themselves facing overwhelming human need. The poor began turning to the church, not a government office, for aid. Alarmed about the sudden increase of homeless people in his city, the mayor of New York made a creative proposal to church leaders. Thirty-six thousand people wander New York streets without shelter, he said; if each of the city's 3,500 churches and synagogues would care for ten of them, they would solve the problem of homelessness. The mayor brought to urgent attention a chronic pain that plagued a large city.

Churches responded defensively. One Protestant leader seemed of-fended that he had first read of the proposal in the newspaper. "It is a very complex situation and the remedy will be complex," said another. "There are many problems of implementation." Most asked for time to evaluate the proposal. They claimed their houses of worship were ill-equipped to shelter the homeless. Only seven congregations responded affirmatively.

Although the mayor's proposal indeed had a complex dimension, its simple appeal to charity stands in direct line with the message of the Old Testament prophets, Jesus, and the apostles. "Share your bread with the hungry and bring the homeless poor into your house," said Isaiah (58:7 ESV). In the early church, members routinely brought vegetables, fruit, milk, and honey to distribute to widows, prisoners, and the sick. Fol-lowing their path, modern churches have taken the lead in operating soup kitchens and homeless shelters, so effectively that the US gov-ernment sponsored "charitable choice" legislation to aid their efforts.

In no way do I mean to imply that chronic pain will gradually fade away. No one who has worked in a country such as India could easily come to that conclusion. I think of the crush of refugees fleeing war and violence; and of a lonely woman, abandoned by her husband, left alone to raise children with insufficient resources; and of released prisoners struggling to reenter society; and of monumental problems of health in the developing world. Neither governments nor the church will relieve all that suffering. More important are the attitudes and energy with which we respond to these chronic pains. Do we grow numb and insen-sitive? Do we react with a quick burst of enthusiastic support that wanes over time?

People with chronic pain, such as quadriplegics or the parents of dis-abled children, describe a common pattern: friends and church members initially respond with sympathy and compassion, but over time they lose interest. Most people find an ordeal with no end in sight unsettling and can even come to resent the one who is suffering.

I retain a clear memory from my childhood of the monthly charity of my Aunt Eunice in England. She kept a little book from the Aged Pil-grims' Friend Society and visited women from that list every month without fail. I accompanied her as she brought money or food or clothing or Christmas packages to those elderly women. In her own quiet, unglamorous way, Aunt Eunice taught me how to turn impersonal, chronic pain into a personal experience of sharing. She insisted on visiting the women, not mailing them packages, and she kept up her simple ministrations faithfully for years.

A physical body's health can be measured in large part by its response to pain. Pain management requires a delicate balance between proper sensitivity, to determine its cause and mobilize a response, and enough inner strength to keep the pain from dominating the whole person. For the Body of Christ, the balance is every bit as delicate and as imperative.

FRICTION AND LUBRICATION

Not all chronic pain in the physical body is debilitating. Less intense forms affect as many as one hundred million people in the United States alone. Persistent pain, often located in the knees, hips, or lower back, affects more Americans than diabetes, heart disease, and cancer combined. Joint replacements, stem cell therapy, and vertebrae fusions have become commonplace, offering a more recent solution to one particular kind of chronic pain.

When parts work together closely, they generate friction. I was reminded of this danger when a concert pianist in England consulted me. "I can no longer perform," she told me. "I can't concentrate on the flow of music or the rhythm. Instead, I can only think of the pain that shoots through my hand whenever the thumb moves at a certain angle." She had recently canceled a series of concerts because of that grating pain, even though she retained all her skills of musical interpretation, muscle action, sense of touch, and timing.

The trouble emanated from a small arthritic area between the two wrist bones at the base of her thumb, and I suggested she continue to play in a way that moved that joint minimally. "But how can I think about Chopin when I have to worry about the angle of my thumb?" she protested. Each time she started to play, her attention riveted on the painful friction of that one roughened little joint.

Treating patients such as this pianist prompted me to study the type of lubrication in our joints, and I gained a new appreciation for how healthy joints work so smoothly without pain. At the Cavendish Laboratory in Cambridge, England, a team of chemists and engineers were seeking a material suitable for use in artificial joints. They found that a joint such as the knee has only one-fifth the friction of highly polished metal—about the same friction as ice on ice. *How is this possible?* they wondered.

Further research revealed that joint cartilage is filled with tiny channels bathed in synovial fluid. As a joint moves, the part of the cartilage bearing the strain compresses, causing jets of fluid to squirt out from these canaliculi. The fluid forms a sort of pressure lubrication that lifts the two surfaces apart. When the joint continues to move, a different part of the surface bears the stress; fluid in the new area squirts out while the area just relieved of pressure sucks its fluid in. Thus, in active movement the joint surfaces do not really touch, rather they float on jets of fluid. The Cavendish engineers were astonished, for boundary lubrication and pressure lubrication were recent inventions—they had thought.

Considering how often joints and bearings require attention in a machine, my joints amaze me with their ability to last for decades without squeaking or grinding. Even so, despite their remarkable powers of lubrication, the body's joints can deteriorate as their gliding surfaces begin to wear thin. As I age, my joints have begun to ache and throb, a natural response to years of wear.

Rheumatoid arthritis, an autoimmune disorder, poses a far more serious problem. Suddenly the body's immune system turns cannibalistic, mistakenly attacking the joints as if they were foreign substances. The synovium thickens and inflames, and a civil war breaks out. The defense mechanism itself becomes the disease.

I see clear parallels in the spiritual Body. Its "joints" are those areas of potential friction where people work together in some stressful activity. A form of spiritual rheumatoid arthritis sometimes attacks individuals who are doing good and important work. Members become hypersensitive, taking offense at imagined criticism or even a disagreement over politics or theology. Their own dignity and position become more important than the harmony of the group.

Some may assume Christians are less susceptible to friction because of the ideals and goals they hold in common. In fact, Christian work can increase friction as the pressure to "be spiritual" exacerbates working tensions. At the Christian Medical College in India we had a psychiatrist who counseled many missionary clients. Highly motivated, working in lonely places, often with just one colleague, missionaries fall prey to acute personal tensions. Friction may result from something as trivial as an ill-timed joke, a tendency to snore, or the way a roommate picks her teeth.

When I experience friction with colleagues or fellow church members, I have to ask myself whether it stems from my own pride or righteous indignation. Could my irritation be causing more harm than whatever I am irritated about? Sometimes the grace of God comes in the form of little squirts of synovial fluid that helps older Christians to get along with the young, who have different notions of proper behavior and appearance—and that also helps the young to understand what it must be like to live with brittle, worn-down cartilage.

The human body goes to remarkable lengths to prevent friction, and the Body of Christ should learn from it the need to lubricate possible friction areas as we cooperate in mutual activity. It takes little grace to get along with people who see eye to eye with me. Grace is put to the test when I work together with people who have different styles and see the world differently—who "rub me the wrong way."

PART SIX

The BODY'S CEO

*From the brain, and from the brain only, arise
our pleasures, joys, laughter and jests, as well
as our sorrows, pains, griefs and tears.*

HIPPOCRATES

CHAPTER NINETEEN

BRAIN

The Enchanted Loom

WHAT FORCE RACES THROUGH THE BODY to connect its many parts? Could it possibly be electricity? To former generations, the very concept of electricity was as mysterious, and terrifying, as nuclear energy is to ours. Benjamin Franklin risked his life by launching a kite into the teeth of that fiery power. What relevance could the feared juice of the heavens have to nerve cells buried in the body's soft tissue?

Before Luigi Galvani, an Italian who lived thirty years after Franklin, scientists and doctors had accepted the ideas of the Greek physician Galen, who described the body's communication system as a flow of *pneuma,* or spirit, through a network of hollow tubes. Then one humid day, Galvani brought a few frogs home for dinner and hung them on his porch.

Following one of those farfetched hunches that have formed the history of science, he beheaded the frogs, skinned them, and ran a wire from a lightning rod to the frogs' exposed spinal cords. He recorded what happened next as a summer thunderstorm swept across Bologna: "As the lightning broke out, at the same moment all the muscles fell into violent and multiple contractions, so that, just as does the splendor and flash of the lightning, so too did the muscular motions and contractions . . . precede the thunders and, as it were, warn of them."

Galvani did not bother to describe the expressions on the faces of his guests, who watched headless frogs jerk and twitch as though kicking across a pond. He stuck to the science, concluding that electricity, not pneuma, had surged through the nerves of the frogs and stimulated movement in dead animals. Entranced, Galvani performed many other experiments. One bright day he hung several beheaded frogs on copper hooks just above the iron railing of his porch. Whenever one of the frog legs drifted toward the railing and made contact, it jerked violently. Reflexes during a lightning storm are one thing, but dead frogs high-kicking on a sunny day—that's the kind of discovery to set the scientific community on its ears. And so it did.

Galvani's rival, Alessandro Volta, decided that the electric current had nothing to do with the frogs and everything to do with two dissimilar metals joined by an organic conductor. He went on to invent the battery, and we have him to thank for flashlights, laptop computers, and cars that start on below-zero mornings. Galvani insisted the reaction came from "animal electricity," and we have him to thank for EKG monitors, biofeedback machines, and electric shock treatment.

WIRES WITHIN US

The neuron plays the key role in carrying out orders from the head. Inside each of us, twelve billion neurons, so fine that a hair-width bundle of them contains one hundred thousand separate "wires," lie poised for action. Medical specialists view them as the most significant and interesting cells in the entire body.

The neuron begins with a maze of minute, lacy extensions called dendrites, which, like the root-hairs of a tree, ascend to a single shaft. These dendrites wrap around every square millimeter of skin, every muscle, every blood vessel, and every bone, interweaving so intricately that even through a microscope it is nearly impossible to discern where one ends and another begins. I liken the sight to standing on the edge of a forest on a winter day. Before me marches a line of several hundred trees, each thrusting dark lengths of snow-laced branches up and out. If all those trees were somehow compressed into a few square yards, with their branchlets filling in the spaces without touching each other, the resulting image would resemble a nerve bunch in the body.

A debate raged in neurophysiology for decades: Do the dendrites actually touch? In the electrical wiring of a home or the circuit board of a computer, every live wire connects with every other wire, resulting in a closed loop. Eventually, it became clear that in the human body each of the twelve billion neurons stops just short of its neighbors, forming a precise gap called a synapse.

The synapse allows for staggering complexity. Take just one motor neuron controlling one muscle fiber in the hand. Along the length of that single nerve cell, knobs from other neurons form synapses at many junctions. A large motor nerve may have ten thousand points of contact, and a brain neuron may have as many as eighty thousand. If an impulse prods one motor nerve into action, instantly thousands of other nerve cells in the vicinity go on alert.

I want to move my index finger to type a letter, so my brain sends a command to a motor nerve. That nerve relies on surrounding neurons to help it compute how many muscle fibers should be mobilized for action, as well as which opposing muscles to inhibit. Neurons carry these electrical messages, as many as a thousand per second, with an appropriate pause between each. Every impulse gets monitored and influenced by all ten thousand synaptic connections along the path. A stupendous crackling wildness streams through all of us at every moment.

MASTER OF DELEGATION

The brain does not consciously order every decision in the body, for that would defy the management principle of delegation. Instead, a dependable reflex system handles many ordinary situations.

Earlier I referred to the knee-jerk reflex, which doctors often test. Even when I tell a patient to stifle the reflex, the leg recoils anyway. I smile with approval, for I value the reflex as a sign of protection, not insubordination. Normally—in fact, almost always except in the case of a reflex test—abrupt tension in that tendon means a person's knee has just absorbed a sudden stress, and the reflex straightens the leg to avert a fall.

The brain delegates such safeguards to the reflex arc in the spinal cord, which explains why my patient cannot easily overrule the reflex kick.

It shows good management, this delegation to sneeze, cough, swallow, salivate, and blink. *Blink:* we do it without thinking, some twenty-eight thousand times per day. I yearn for such a reflex in my leprosy patients, many of whom go blind because their deadened pain cells do not inform them when the dry cornea needs a lubricating blink. We can forestall blindness in many patients by teaching them to blink with regularity, and I naively assumed that patients with eyesight at stake would make eager learners. I soon discovered that conscious movements are not nearly as reliable as reflexes.

We have to train our leprosy patients to blink, using placards and stopwatches—drilling, scolding, praising, and cajoling them. The higher brain resists giving priority to something as elementary as a reflex (who would force a supercomputer to count to ten every thirty seconds?). Many patients do not learn, and their eyes eventually dry out.

Some functions need more direction than the robot-like response of reflex. The brain stem itself coordinates the next level of guidance, the subconscious regulators of breathing, digestion, temperature, and circulation. These need more attention than reflexes: when I race up the stairs, my heart and lungs must shift into another gear, and the act of breathing alone must coordinate ninety different chest muscles.

Highest of all in the hierarchy of the nervous system are the cerebral hemispheres of the brain, the body's CPU, most shielded by bone and most vulnerable to injury if that barrier is ever breached. There, some eighty billion nerve cells and many more glia cells (the biological batteries for brain activity) float in a jellied mass, sifting through information, storing memories, creating consciousness. Anatomists estimate the human brain contains more than a hundred trillion synapses. From all this buzzing activity, we make conscious choices.

I marvel at the nervous system's sophisticated design. When a sudden danger—touching a hot stove, protecting eyes in a dust storm—requires a quick response, the brain delegates it to a reflex loop that functions below the level of consciousness. Yet the body reserves the right to overrule this reflex loop under unusual circumstances. An expert rock climber clinging to a precipice will not straighten his leg when a falling stone hits the patellar tendon; a society matron will not drop a too-hot cup of tea served in Wedgewood china; a mother does not reach out to break her fall when holding a baby.

THE FINAL COMMON PATH

Although the hierarchy seems neatly ordered, one anomaly keeps showing up. The final decision, the localized "will" that controls muscles and movement, resides not in the magnificent crevasses of the brain but in the humble neuron that controls the muscle fibers. Sir Charles Sherrington discovered this discomfiting feature and grandly labeled it "the final common path."

Along its length, each neuron receives a spray of impulses from surrounding nerves. It stays alert to muscle tension, the presence of pain, the action of opposing muscles, the degree of strength required for any given activity, the frequency of stimuli, the oxygen available, the fatigue factor. An order from the brain arrives: lift a heavy box. Accomplishing that act will involve a whole army of motor units. After sifting through all the synaptic signals, the motor neuron itself decides whether to contract or relax its particular muscle fiber. After all, it is best equipped for such a decision, being in intimate contact with so many local synapses.

Only the "final common path" can decide between incompatible commands and reflexes, and we should be grateful. I stand on a cliff on one of the sheer granite hulks in the Rocky Mountains. Ahead of me, just beyond my reach, I see a delightful wildflower I cannot readily identity. I plant my feet, squat down, and lean forward, adjusting my camera under instructions from my brain. My close-up lens approaches within inches of the wildflower when suddenly a string is jerked and, like a marionette, I tip backward away from the flower. My heart is pounding, and I look around to see who interrupted my photography. No one is there, save a raucous, scolding jay.

Ever since I peered over the edge of the cliff to the ravine two thousand feet below, my cells have been chemically flooded with a heightened awareness of the potential danger. My conscious brain wanted a picture of the flower, but my subconscious reflexes received alarming reports from the balance organs of my inner ear. Short-circuiting the conscious brain, an "Emergency!" message convinced the nerve cells that control my muscles to yank me backward.

Sometimes my brain overrules, and sometimes it delegates. The response to its commands ultimately depends on the local, autonomous cell—the final common path. The microscopic computer in each nerve cell gauges my intentions, consults other muscles, calculates available energy, and monitors any signs of pain; only then does it fire a yes or no order to its assigned muscle group.

THE BRAIN EXPOSED

Medicine knows no more daunting procedure than brain surgery. No one who opens a human skull can avoid a grim sense of defilement, almost like sacrilege.

I have observed a living human brain on maybe a half-dozen occasions. Each time I feel humble and inadequate, an intruder. Who am I to invade the hallowed place where a person resides? Solzhenitsyn once referred to a man's eyes as "sky-blue circles with black holes in the center and behind them the whole astounding world of an individual human being."

During my medical training, I chose as my senior project the task of exposing the major nerves in the head, tracking them all the way into the brain. Two years of medical school had not steeled me sufficiently for the experience of getting my own cadaver head, whole and perfectly preserved though shrunken slightly by the chemicals. It had belonged to a middle-aged man with plentiful hair and bushy eyebrows.

Over the next few weeks I spent most of my waking hours with the head of my anonymous friend. "That skull had a tongue in it, and could sing once," Shakespeare wrote, and I had to push from my mind images of this hunk of wrinkled tissue on the table singing, talking, winking, smiling. I felt almost grateful for the pungent odor of formaldehyde that seeped through my skin and affected the taste of food, for it reminded me I was carving away not on a man but on a specimen of preserved tissue.

The skull, I learned, is a nearly impregnable orb. In my training I had watched as brain surgeons huffed and puffed, leaning in heavily to force a whirring drill bit through its quarter-inch armor. The bone barrier had sealed off my cadaver's brain from all direct encounter with the outside world. Paradoxically, that secluded brain had stored all its owner's knowledge of the world, thanks to the frail, white nerves leading into it. One nerve had controlled all the subtle movements of his lips that made possible speech and eating and kissing. Another had brought in every nuance of color and light to form his visual construct of the world.

I began my exploration with the familiar shapes of eye and ear. Then I proceeded inward, like an explorer searching for the source of the Nile, following a small strand of white into the penetralia of the brain itself. I chiseled the facial skeleton in thin layers, coaxing out slivers of bone while taking care not to cut too deep and sever the nerve. The orbit of the eye, for instance, consists of seven bones fused together in a socket. Fortunately, I had worked as a stone mason for a full year, and before long, the process of chiseling away layers of bone the thickness of tissue paper seemed natural and even artistic.

I remember being most impressed by the range of textures. I picked up a scalpel to make a slice through satiny muscle and fat, holding my breath and keeping the blunt edge toward the nerve—one quiver of my finger would sever it. Next, I laid down the scalpel, picked up a mallet and chisel, and attacked the bone with all my force.

After several arduous weeks of dissecting, fine white fibers led from the cadaver's ear, eye, tongue, nose, larynx, and facial muscles to disappear into the cavity that contained the brain. Finally, I was ready to enter the brain itself. After vigorously sawing through bone, I reached the three membranes, or meninges, that sheathed the brain. I slit each one, remembering with a smile the arcane Latin names I had learned in anatomy class: *dura mater* (hard mother), *arachnoid* (cobweb), and *pia mater* (tender mother).

The innermost membrane fit like Saran wrap over the convolutions of the brain, and when I punctured it, a small piece of the brain bulged through the opening like a tiny fist. I stared at it a full five minutes before continuing. The organ weighed barely three pounds, yet that fragile, grayish jelly once contained a person's entire life experience.

I touched it: gray matter had the consistency of cream cheese, softer than any bodily tissue I had yet encountered. Its landscape dipped and rose and turned in on itself—a topographical map of all the mountains on earth compressed into a small space. Red and blue lines crisscrossed the topography, and I breathed a prayer of thanks that I was practicing on a dead brain. A surgeon operating on a living patient spends much time avoiding those vital channels of blood and stanching the vessels cut by his scalpel.

Although I had hoped to trace sensory nerves to their origin, the brain does not easily yield to map-making. Nerves there, ensconced in the ample armor of skull, have a doughy consistency and will break at the slightest tug. And even the most experienced surgeon has difficulty with orientation in the brain, for everything appears soft and white, like an Arctic landscape.

My professor took tutorial delight in my project, displaying the pickled result in a medical museum. I had schoolboy fantasies of becoming a brain surgeon. Years later, when of necessity I attempted a few hazardous ventures into neurosurgery, I felt immensely grateful that I had *not* pursued that enormously challenging field.

FORCES WITHIN FORCES

Brain surgery on a living patient is a very different experience. Sometimes, the patient stays awake in order to cooperate with the exploring surgeon, and the fact of the patient's consciousness serves to dampen the normal prattle of surgery. Standing on the sidelines of such procedures, I notice the sounds: the faint electronic beeps of monitoring machines, the sighs of the respirator, the shrill whine of the drill, slight pops from the electric cautery, the clinking of instruments being passed around like dinnerware. The object of all this attention glistens in the bright lights, and if I look closely I can see it heaving gently. The brain is alive.

Brain surgery remained in a primitive state until one remarkable discovery. When a surgeon inserts a needle-like electrode into a portion of the brain and switches on the current, the brain responds, indicating what function that area controls. The patient will say something like "I feel a tingling sensation in my left leg" if the surgeon lightly stimulates a particular spot.

Wilder Penfield, a neurosurgeon in Montreal, recorded remarkable results from such exploration. While trying to locate the source of epileptic seizures, he found that his electrical stimulations could evoke specific memories in sharp detail. One young South African patient began laughing, reliving second by second an incident on a farm in his native land. A woman recalled every note in a symphony concert she had heard long before. The memories revived in such detail for one patient that she recollected a scene at a train crossing years before, describing each train car as it went by. Another counted aloud the number of teeth of a comb used in childhood.

Thanks to these discoveries and more recent results from functional MRIs, we now have a fairly reliable map of the brain. Most brain research centers on the cerebral cortex, which is far more advanced in humans than in any animal. The thickness of the sole of a shoe, the cortex directs the higher activities of thought and memory while also processing all the information received from the sensory organs. The majority of neurons live in that layer of gray matter, the fertile topsoil of the brain.

Nerve cells divide into two groups: "the way in," or *afferent* cells that carry impulses from the body into the brain, and "the way out," *efferent* cells that carry instructions from the brain out to the extremities. All visual images, all sounds, all touch and pain sensations, all smells, the sensations of hunger, thirst, and sex drives, muscular tension—all the noise from the entire body—occupy only one in a thousand of the brain's cells, the afferent cells. Similarly, the efferent cells make up a fraction of one percent of brain cells, enough to control motor activities: playing a musical instrument, speaking a language, dancing a ballet, typing a letter, playing a video game.

The remaining, vast majority of brain neurons join together in a network of intercommunication to allow the processes we know as thought and free will. One brain biologist likens this network to billions of bureaucrats constantly phoning each other about plans and instructions for keeping a country running. More poetically, Sir Charles Sherrington rhapsodized about an "enchanted loom" with lights that flash on and off as messages weave their way through the brain—the very image later captured by functional MRIs.

The entire mental process comes down to the brain's cells spitting chemicals at each other across synapses. Its complexity defies description, with the total number of connections far exceeding the number of galaxies in the universe. A mere gram of brain tissue may contain as many as four hundred billion synaptic junctions. As a result, each cell can communicate with every other cell at lightning speed—as if a population far larger than earth's were linked together so that all inhabitants could talk at once.

Mercifully, we are hardly aware of the process. I decide to write the next sentence; in a flash my brain computes first the thoughts and then the words I will use, then the elaborate coordination of muscles, tendons, and bones required to type the words. Before I finish typing, my brain begins composing the sentence to follow.

Steven Levy records what happened when, on a visit to Princeton, he came across a jar containing Albert Einstein's brain:

> I had risen up to look into the jar, but now I was sunk in my chair, speechless. My eyes were fixed upon that jar as I tried to comprehend that these pieces of gunk bobbing up and down had caused a revolution in physics and quite possibly changed the course of civilization. There it was.

In the human head, concludes Nobel laureate Roger Sperry, "there are forces within forces within forces, as in no other cubic half-foot of the universe that we know." Perhaps if I worked on brains daily I would grow more callous. I doubt it—the neurosurgeons I know still speak of the brain in hushed, almost worshipful tones.

And yet nothing on earth is so fragile. One dosage of a powerful drug can permanently upset the delicate balance inside a brain. One bullet may destroy it or one spill from a motorcycle. Deprived of oxygen for a mere five minutes, the brain will die, and with it the whole body.

CHAPTER TWENTY

IMAGE RESTORED

EVEN ON A QUIET DAY my brain vibrates with activity, processing five trillion chemical operations every second. I am most conscious of the traditional five senses: sight, hearing, touch, taste, and smell. Other reports are more subtle: I know intuitively the tilt of my head, the bend of my elbow, the position of my left foot. More sensors inform me of the need to stop for lunch; my stomach "feels empty." My brain, isolated in its thick ivory box, receives all these reports in a kind of electrical Morse code.

Even so, the brain can uncannily reproduce reality. Consider a Beethoven piano sonata. Deaf in his later years, Beethoven never heard the music he composed—that is, the eardrum, three bones, and sound receptor cells never participated in the creative act. Yet his brain internally reconstructed tone and harmony and rhythm so that he did "hear" it. No molecules danced their tarantella; the music took form in silence, cerebrally, in code.

Today, if my musically talented wife picks up a written score of Beethoven's *Sonata Pathétique,* she identifies it almost at once. She can hum along as she reads, relying on her own mental bank of sounds to hear it in her mind. And if a radio station happens to broadcast a performance, she recognizes it after hearing a few measures. How many billions of computations are required for the mind to recollect a piece of music—a feat performed at blinding speed with little conscious effort?

As you read this paragraph, you hardly note the individual letters forming each word. You do not spell them out one by one, reassemble them into a composite, and scan a dictionary for the meaning—though in reality your mind does all that, subconsciously. It works so fast that when I speak, using letters, words, grammar, and punctuation, I concentrate solely on the meaning of what I want to communicate. Neurons with stored knowledge freely supply the individual elements, and my central nervous system arranges the glottal puffs and slides to create intelligible sounds.

The writer Jeanne Murray Walker describes the result: "Words are only scraps of sound, rags of wind, bits of vibration we shape like music with our tongues and teeth and breath. But if you're reading this, words are where the two of us are meeting. Words help us grasp one another, they nerve us to go on."

My brain presents the world to me not in reductionist blips of data but wholly, conceptually, meaningfully. And herein exists a great mystery. The mind that coordinates all this profound activity lies locked away. The brain itself never sees: if I expose one to bright light, I risk harming it. It never hears: the brain is so sheltered and cushioned that it feels only the most reverberant sensations. The brain does not experience touch: it has no touch or pain cells. Its temperature varies no more than a few degrees; it has never felt hot or cold.

My perception of the world forms from millions of remote stations sending reports in a digital code to a bony box that has never directly experienced those sensations. The taste of chocolate, the prick of a pin, the sound of a violin, a view of the Grand Canyon, the smell of vinegar—all these reach my consciousness via impulses that are virtually identical. I perceive the outside world because tiny, flower-shaped neurons have shot chemicals at each other.

The mystery of the person I am, Paul Brand, centers in the brain. Every other cell in the body expires and gets replaced but not the neurons. How could we function if memory and knowledge periodically sloughed off like skin cells? In terms of my physical body, I am a different person from my younger self—the exception being my long-lasting nerve cells. These maintain the continuity of selfhood that keeps the entity of Paul Brand alive.

From the darkness and loneliness of that bony box, my brain reaches out to reality with millions of living wires. They extend like tendrils of a plant, stretching hungrily toward stimuli from the world beyond.

THE DESCENT

I have described the brain's hierarchy in detail because from it we can learn about leadership. The brain delegates its tasks to the brain stem, the reflex arc, all the way down to the final common path of individual neurons. This network of cooperation, involving every cell in the body, makes possible amazing feats of genius and athleticism. Always, though, the brain reserves the "higher" activities of cognition and prioritizing for itself.

The most effective organizations of people follow a similar pattern of delegation and mutual interdependence. I have extended the analogy in order to lay the groundwork for spiritual lessons we can learn from the physical body. As a Christian, I turn to the analogy of Christ as Head of the church, a title the New Testament applies to him seven times. Insights from the brain shed light on the style in which headship is exercised.

The body hints at a fundamental principle of how spirit, or mind, interacts with matter. God, a Spirit unbound by space and time, in an act of deep humility took on the confinement of matter and time—an event that Christmas celebrates. "'Twas much, that man was made like God before / But, that God should be made like man, much more," wrote John Donne. The actual incarnation, however, spanned only thirty-three years.

From the outset Jesus predicted his departure, foreshadowing a time when he would leave the work in the hands of his followers. After his departure, Jesus Christ receded to the role of Head in order to create a new Body, this one composed not of living cells but of men and women from all over the world. "As you sent me into the world," Jesus reported to his Father, "I have sent them into the world" (John 17:18). Can the shift be expressed more succinctly?

In one sense Jesus' departure from earth was an ascension—church calendars call it that—though in another sense it was a further condescension. God elected to make God's presence known through people like us—not in one body but in many, not in one perfect Son but in millions of ornery children of all races, sizes, IQs, personalities, and genetic traits. The Spirit has chosen to make our prayers, our compassion, our actions, our proclamation of truth and justice a primary means of relating to the world of matter.

Today, we are God's medium, Christ's Body. When you look at me, you don't see the whole Paul Brand; rather you see a thin layer of skin cells stretched across my frame. The real Paul Brand resides inside, especially centered in my brain, hidden from the outside world. Even more so, we cannot "see" God; we lack adequate perceiving organs. Rather, God becomes visible through the members of the Body.

We are called to bear God's image corporately because any one of us taken individually would present an incomplete image, one partly false and always distorted. Yet collectively, in all our diversity, we can come together as a community of believers to restore the image of God in the world.

This style of involvement with the created world—power exercised "from below" rather than from above—raises urgent questions. For the agnostic, such questions take on a tone of accusation: "If there is a God, let him prove it somehow! Let him step in with divine power and straighten out the mess of this world." As a Christian, I struggle not so much with the question *Is God really there?* as with *Why has God chosen such an indirect and hidden way of working on earth? Why rely so heavily on unreliable human beings?*

I get a clue into one possible answer whenever I take on a teaching assignment and experience the peculiar satisfaction of work done through others. If I were to calculate the number of hands that I personally operated on, I would likely come up with a number around ten thousand. That number seems large, a testament to my advancing age. However, as I reflect further, I realize how negligible that number is. Millions of people in the world suffer from leprosy, a quarter of whom have hand damage. In a lifetime of surgery, putting in as many hours as I can muster, I have personally helped only a tiny fraction of those with needs.

Many times, though, I have visited a tiny rural clinic in a place like Borneo and watched a young doctor perform procedures that derive from those we developed at Vellore. In Pakistan, South Korea, Ethiopia, and virtually anywhere leprosy work thrives, you will find students who were trained at Vellore or Carville. Nothing, absolutely nothing, fills me with more joy than to see the seeds of what I taught now sprouting in others' lives. What I invest in a classroom can multiply a hundred times what I could possibly achieve on my own.

That realization gives me insight into God's way of working in the world. Just as a teacher extends his or her work through students, and a brain expresses itself through loyal cells, God expresses God's own self through a Body in which Christ serves as Head. During his time on earth, Jesus had no influence on the three places I have lived: Britain, India, and North America. In the centuries since, however, his followers have established loyal outposts of the kingdom of God in those places and many more, just as he commanded.

"Whoever listens to you listens to me; whoever rejects you rejects me," Jesus once told his followers (Luke 10:16). The identification of the Body with its Head is that complete. A little later, on the night of his arrest, Christ explained his imminent death for the confused and somber disciples. "It is for your good that I am going away," he said (John 16:7). Although they did not know it at the time, the new era of headship was underway.

THE RISK OF DELEGATING

Dorothy Sayers names three great humiliations God has willingly undergone. In the first, the incarnation, God stripped off the prerogatives of deity and descended to live as a human being on earth. In the second, the crucifixion, God's Son suffered an ignominious death. The third humiliation, she says, is the church. God in the person of Jesus Christ is one thing, and God in us is quite another.

The Head working through member cells involves a sort of abdication in which God sets aside omnipotence and adopts an invisible, behind-the-scenes role in human history. In so doing, God riskily entrusts the divine name and reputation to imperfect human beings. Members of Christ's Body have sullied God's reputation by such misdeeds as launching crusades, torturing heretics, and trafficking in slavery. The flaw is not in the Head, to be sure, but the humiliation is there.

I turn again to the analogy of the human body. A healthy body relies on proper channels from the brain to body parts, as well as a commitment from individual cells to do the will of the head. Persons afflicted with neurological diseases live with the constant frustration of disobedient cells. Some of them possess magnificent minds, such as Stephen Hawking, the late physicist. And yet many so afflicted are judged ignorant or mentally deficient because of the disruption between mind and body.

In a spastic disease or paralysis, somewhere, often in the descending fibers to the cells, communication breaks down. A paraplegic can lie in bed and plot how to move her toe, then *will* it to move with her full mental energy, but because of a broken connection the toe will not move.

In the spiritual Body, a cell must submit to orders from the Head, for only the Head can judge the needs of the whole Body. Obedience alone determines an individual cell's usefulness in Christ's Body. Errors inevitably creep in—spastic movements, if you will. It cannot be easy for the omnipotent One to endure the humiliation we bring on. (Is there a divine counterpart to the frustration a paraplegic feels?)

I must quickly add that analogies to the spiritual Body apply only partially, for dysfunction there never results from brain damage. But many nerve disorders—cerebral palsy, for example—occur when synaptic channels below the level of the brain somehow clog up. Neurologists occasionally encounter the unusual condition called alien hand syndrome. These patients have hands that seem locked in a tug of war: one hand may spontaneously seize objects from the other hand or try to restrain the other's movements.

The apostle Paul, master of metaphor, gives a precise description of a person suffering from this sort of disconnection in his letter to the Colossians. The person he describes has focused on judging neighboring cells rather than on obeying his own orders from the Head.

Such a person . . . goes into great detail about what they have seen; they are puffed up with idle notions by their unspiritual mind. They have lost connection with the head, from whom the whole body, supported and held together by its ligaments and sinews, grows as God causes it to grow. (Colossians 2:18-19)

REPATTERNING

The brain assigns a specific region to govern each finger, each toe, each significant body part. For instance, my brain devotes a site to all the associations of my right ring finger. How is it used in playing the guitar? Does it steady my hand in writing? Does it bear a scar from a prior injury? The brain stores these memories and abilities. If my finger is used prominently, as in playing the guitar, the brain will have an increasing richness of association with the finger.

In my surgical practice I sometimes disrupt these associative pathways and try to establish new ones. For cosmetic purposes we give some leprosy patients new eyebrows by cutting a swatch of hairy scalp and tunneling it under the forehead to the eyebrow area. Still attached to the scalp's original nerve and blood supply, the patient's new eyebrow "feels like" part of his scalp. If a fly crawls across the transplanted eyebrow, the patient will likely respond by slapping his crown.

Or in a tendon transfer procedure I may move a healthy tendon from the ring finger to replace a weak or useless one on the thumb. "Move your thumb," I instruct the recuperating patient, and nothing happens. The patient just stares at her hand. "Now move your ring finger," I say, and the thumb springs forward. Over time the patient must repattern her brain to transfer the sensation of ring finger motion to the thumb. It can take months to reestablish smooth patterns, and many patients over the age of forty never fully adapt.

The image of thumb cells struggling to receive a strange new set of orders from the head helps me appreciate the apostle Paul's injunction to be transformed "by the renewing of your mind." Elsewhere, Paul exhorts the Philippians to "Let this *mind* be in you, which was also in Christ Jesus" (Philippians 2:5 KJV, emphasis added). I think of spiritual disciplines as the repatterning therapy required to develop a smooth, steady stream of transmission between a cell and its Head.

Some Christians are now rediscovering ancient disciplines of faith. Through meditation, fasting, prayer, simple living, worship, and celebration, we can develop an ever-increasing richness of association between ourselves and the Head. The simple practice of repeating prayers from the Book of Common Prayer can help discipline the mind. In neurophysiology as well as spirituality, repeated acts of obedience strengthen the connections. A good pianist does not consciously think through the motions involved for each finger to strike a note; the fingers follow pathways laid down through hours of practice.

For the beginning Christian, the process of learning Christ's mind may seem mechanical and ungainly. The Christian walk, like the toddler's walk, begins with false starts and stumbles and missteps. Gradually, though, the muscles and joints in knee and leg and foot learn to cooperate together so that the child runs across the room without giving conscious thought to the process. Each new skill will begin in a fumbling, error-prone way until movements become fluid and natural.

Of all the marvelous aspects of the human body, I know of no greater wonder than that every one of the trillions of cells in my body has access to the brain. And in the spiritual Body, I know of no greater wonder than that each one of us has direct contact with the Head. Incredibly, it seems that God yearns for contact with the disparate members of the Body. God listens to our input, considers our requests, and quite literally uses that information to influence activities in the world. "The prayer of a righteous person is powerful and effective" (James 5:16).

LIVING BY INSTINCT

Let me shift the image from physiology to nature. On my walks outdoors I have been watching a pair of young orchard orioles build their first nest. A few branches away hangs a nest built last year by older orioles, a nest so sturdy that it survived a winter storm that tore branches from the tree. Yet these young birds never fly over to inspect the old nest or study it for design innovations. They know exactly what to do. They neglect eating in the urgency of their task.

The orioles begin by selecting the best location. They look for a small branch with a well-spaced fork on which to weave the nest. The branch must be so thin as to droop a little from the mere weight of the leaves, in order to guard against squirrels. Foliage must surround the site to conceal their offspring from hawks and other predators flying high above.

Once they agree on a prime location, the birds search for individual blades of grass, of one type only, that conform to a certain length and consistency. One of the birds stands with a foot perched on each twig of the fork, holding a blade of grass under one foot. Using only its beak, it ties a half-hitch knot around that twig, leaving a long end to dangle. After flying away for another blade, it ties a half-hitch on the other twig and then weaves the two blades together. It repeats this process over and over, plaiting the strands of grass into a thick cable. The nest itself will swing between these cables. After several days of selecting, weaving, plaiting, and wattling, the two birds will have a neat, spherical home, strong enough to withstand gale-force winds.

Inside my house, my wife is knitting me a pullover sweater. As I watch the birds, I can see her through a window. Her skein of wool is the product of the skill and experience of shepherds and shearers and spinners and dyers. Margaret keeps glancing down at a printed pattern that reflects the artistry and calculations of master knitters. Reading those instructions requires education, and following them employs a skill she has learned over the years. A sweater that I will wear with pride finally emerges, a result of the shared intelligence of many brains over many years. Given a live sheep and told to create a sweater with no outside help, my wife would surely fail.

I know that, even with full concentration and the manual dexterity developed over years of surgery, I cannot weave strands of grass into a hollow globe that will cling to a branch in a storm. I tried it once, and the final product fell apart, limp and useless. Yet I have ten fingers while an oriole has only a beak and two feet. Instinct is the key.

A kindred species, the bunting, follows another imprinted code that guides it across the Gulf of Mexico to a new home five hundred miles away. I have watched migrating birds depart. They sit on reeds in the swamp and look out across an expanse of water that must appear endless. Unencumbered by reason, the birds always head south. Their instinctive wisdom predates the egg from which they hatched.

I sometimes think of the bunting and the oriole as I struggle with spiritual decisions. Messages from God reach me along crowded pathways of communication. With my reason, even while contemplating what the Bible says, I can easily rationalize my way to other conclusions. The commands are hard; they require love and sacrifice and compassion and purity, whereas I concoct excuses that make obedience seem unattainable. At such times, as my own selfishness and pride rear up, I need a force more dependable than reason.

Just such a force comes built into each of us: the conscience or subconscious, a law written in our hearts. We can encourage and mature this instinct by the disciplines of faith. As one example, if I must decide whether to tell the truth in the face of every situation, my life will be hopelessly complicated. But if I have developed a reflex of truthfulness, I can learn to walk as a Christian without having to think about each individual step.

When the crucial moment of choice arrives, I often have little time for conscious reflection. All that has gone before enters into that moment. I think of those tiny birds, the oriole and the bunting, and ask that in renewing my mind God would imprint instructions into me as if they were put there genetically. I ask for an uninterrupted flow of messages from the Head and for a response of faithful obedience.

CHAPTER TWENTY-ONE

LEVELS OF GUIDANCE

IN A HEALTHY PERSON, the components of the nervous system work together in a beautiful, singing harmony. As I walk to work in the morning, I think about my patients and notice the birds perched in the branches; the motor units in my legs need no conscious direction from the brain. Loyal neurons will slow my pace if my heart complains and will take reflexive action if I stumble. Every level in the hierarchy contributes to my well-being.

I dare not leave out one more influence on cellular behavior: hormones. These chemical compounds act as agents under orders of the head to orchestrate major events. Whereas the brain sends precise, targeted messages to specific nerves, hormones issue a general summons that can reach every cell in the body. Although all cells hear the call, only the appropriate ones respond.

I see the power of hormones most dramatically in the phenomenon of pregnancy. The adult woman's body prepares for that possibility every month until one day a fertilized egg settles on the receptive cells of a freshly lined uterus. An alarm sounds, hormones kick in, and the nine-month countdown begins. Molecules that previously have never provoked more than a mild reaction suddenly, after conception, foment revolution.

Progesterone, for example, has visited the uterus monthly in small concentrations, sometimes irritating the lining and causing the noxious reaction of menstrual cramps. Now, after the egg's fertilization, uterine cells fully realize their calling. At once the uterus undertakes a massive reinforcement project, thickening its walls to prepare for the fetus it will soon shelter and protect. Cells pile on top of cells, layering, stretching, dividing, so that eventually the uterine wall grows to a hundred times its previous size.

More and more, the woman's body reorders its priorities toward creating a new life, not just preserving a familiar one. For example, as the body devotes fewer resources to the production of digestive enzymes, nausea may result, the familiar "morning sickness."

A remarkable organ, belonging neither to the mother nor the child, begins to develop: the placenta. From an immunological standpoint, the placenta represents foreign matter to the mother, but her body knows to welcome it. No open passage connects the two beings, no cells cross the membrane, and the mother remains wholly mother and the child wholly child. (Often their respective blood types differ so that any blending of blood might prove fatal.) The placenta forges a supreme bond of symbiotic intimacy.

People who see the placenta only after it has played its role dismiss it with the inglorious name *afterbirth*. In reality the placenta is one of nature's most elegant structures. Burrowing deep into the tissues of the mother, the placenta weaves a lacy web of vessels through a fine membrane so that the chemicals in the mother's blood can diffuse into the child's, and the wastes from the child can be eliminated through the mother. It serves the function of kidney, stomach, and liver for the developing fetus.

Elsewhere, progesterone and its companion estrogen summon responses from cells in remote sites such as the hips, tendons, breasts, and uterine muscles. Ligaments that have always kept the skeleton taut and stable now defy their heritage. They must, for the pelvic bones need to stretch apart enough for a baby's head to pass through. For the mother, the loosening of connective tissue may bring on backaches and a waddling gait known as "the proud walk of pregnancy." Other joints, scanning the very same chemical message, recognize that it does not apply to them, and thus pregnant women are spared the problems of a wobbly head, loose knees, and elbows that dislocate easily.

Hormones that direct cells to loosen hip ligaments and firm up uterine walls also transform the breasts. These molecules have floated past breast cells each day for years, with no effect. At the onset of pregnancy, however, cells that have been relaxing on reserve duty now report to active service. A milk production factory takes shape, with some cells arranging themselves into a tube that branches out through fat tissue, even as the fat cells shrink to accommodate the new ducts. The change in breasts also requires blood vessels to elongate, skin to grow, and pectoral muscles to strengthen.

Finally, after eight or nine months, reports from the womb convince the mother's body that the time for birth is approaching. The mother's blood adjusts its clotting properties to prepare for probable vessel breakage. It also increases in volume by as much as 50 percent, a safeguard against potential heavy loss during childbirth.

Next, the uterus undergoes a burst of contractions and relaxations as extreme as anything the body will experience. Even after delivery, hormones keep flooding the body, in many cases causing the opposite reactions from those in effect just minutes before. The uterus no longer enlarges, it contracts. Blood vessels torn from the placenta seal themselves. The placenta itself, masterful supervisor of much of this activity, exits anticlimactically and gets discarded.

New priorities take over: healing, restoration, and a bonding between two separate beings. In the lovely symbiosis of nursing, the mother now needs the baby just as the baby needs the mother, for the engorged tubes of the breasts must rid themselves of congestion and pain. The placenta's final signals before expulsion, the newborn's first attempts at sucking, and even the baby's cry all combine to stimulate the flow of milk, which the baby quickly learns to ingest. Breast milk communicates two ways: when the baby's spit-back alerts the mother to possible threats, her immune system kicks into gear, which she transfers to the baby through a newly formulated variety of milk.

LISTENING FOR MY ROLE

I look to the physical body for insight into how cells in a spiritual Body can best collaborate. The same chemical messenger instructs uterine cells to contract but cervical cells to relax during childbirth. As one cell among millions joined together in the spiritual Body, how do I discern my specific role, my calling?

I must admit, I sometimes chafe at being one cell. I might prefer being a whole body or another kind of cell with a different role. Gradually I have learned to view myself as a minor part of a great enterprise that will only realize its purpose under God's direction.

In the human body, direct connections to local neurons help inform each individual neuron how to act in community. I believe that God similarly communicates through local agents: spiritual leaders, my loved ones, the community of faith around me. How should the church respond to a decaying inner city? To refugees? To the increasing strains that tear families apart? In the Bible, God lays out principles governing the response of the whole Body and also delegates the details to local groups of Christ-followers.

A sensitive member of the Body will hear many stirring calls to action. Some will describe the desperate plight of people in other lands while others will direct attention to neighbors close by. Some will highlight specific causes such as prison reform, racism, the environment, abortion, poverty, addictions, sexual trafficking; others will stress the contemplative life. All these calls have merit and apply in some degree to each of us. But God's Spirit will instruct us in our *specific* response, and it is to the Spirit that we must listen.

Having spent much of my life in India, where physical needs are great, I rejoice when sensitive Christians are stirred by human needs. Yet the form of our response will vary. Praying for and encouraging those on the frontlines, lobbying for aid, taking a short-term assignment, supporting causes financially—the options are as diverse as the responses of my body's cells to a hormone.

Some Christians seem almost paralyzed over specific questions of guidance. Who should I marry? How should I use my excess income? What role can I play in issues such as racism or world poverty? In short, how do I know God's will for me?

On the opposite end, I have met Christians who use a phrase like "God told me" as a casual manner of speech. "God told me it's time to buy a new car," a person might say, or "I know God wants our church to use our money this way." Actually, I believe that the Bible already contains most of what God wants me to know. A direct hotline from God is not the ordinary way of discerning God's will.

I recall the apostle Paul's phrase from Romans 12: "Do not conform to the pattern of this world, but be transformed by the *renewing of your mind*. Then you will be able to test and approve what God's will is—his good, pleasing and perfect will" (v.2, emphasis added). That passage introduces the first full mention in the New Testament of the body analogy, followed by a list of abrupt commands detailing God's will: Hate what is evil; cling to what is good. Honor one another above yourselves. Share with God's people who are in need. Don't be proud. Live at peace with everyone.

Paul does not dwell on the family dysfunctions and sociological factors that might make such responses difficult. He simply states God's will and admonishes us to "renew our minds"—a process that helps us realign with the hierarchy God has provided. The individual Christian would do better to focus on daily obedience to what God has already made clear rather than seeking some private revelation.

Tellingly, when the New Testament lists leadership qualities or spiritual gifts, it does not emphasize technical skills. For leaders, spiritual qualities are paramount. How committed are they to God? Can they control their own temperaments? What are their families like? The key concern centers on loyalty to the Head. God seems to say, I will work with anyone as long as he or she is committed.

I have learned that the first job of a faithful cell is to listen. I must assess the needs of the Body, peruse the various messages, and wait in readiness. God's Spirit will employ various means to speak to me and instruct me in the way I should go, but only if I tune in. I may want to leap into action, but unless such a response is prompted by the Spirit, it will not help the overall Body.

At times I find myself nearly overwhelmed by busyness, and at such moments I am strongly tempted to shunt aside my normal time with God. Over the years I have learned, with difficulty, that those moments of stress are precisely the times when I most need spiritual renewal. I must carve out a time to bring heaven and earth together. I prayerfully commit my day's clutter to God, asking to see the details of my life in the light of God's will.

GOD'S SENSOR

Mahatma Gandhi, one of the busiest and most famous men in the world, used to set aside Monday as a day of silence. He scheduled no appointments and said nothing all day. He needed the stillness, he said, in order to rest his vocal cords and to promote an inner harmony in his soul amid the turmoil of life around him. I wonder what power would be released if all Christians devoted one day a week to listening to the voice of God in order to discern the coded message for our lives. The Counselor can only direct us if we tune in.

My Grandmother Harris lived to age ninety-four, and I never saw her walk unassisted. Poor health confined her either to bed or to "Grandma's chair" in a quaint room with lace curtains and dark, Victorian furniture. My sister and I would visit that room for about an hour or so each day. Of Huguenot descent, Grandma had us read the French Bible to her so that we could practice the language and also learn the Bible by discussing the passage we had read.

Grandma was bent and wrinkled, and she suffered severe headaches. She rarely laughed and could never comprehend our jokes, yet her quiet joy and peace somehow reached even us play-minded children. We never resented our daily visits to her room. She radiated love.

When Grandma had trouble sleeping, she sometimes lay awake half the night softly reciting chapters from her storehouse of memorized Scripture and praying for her eleven children and scores of grandchildren. My aunts took turns sleeping in her room, and often in the middle of the night Grandma would suddenly call on them to write down her thoughts. She would say, "I sense that Pastor Smith in Ipswich is in need of help just now. Please write to him like this . . ." She would then dictate a letter and ask my aunt to enclose a check.

Days later, when the mail brought a letter of reply, Grandma would beam with joy. Invariably, the letter expressed astonishment that she should have known the precise timing and amount of a need. She would laugh with a pure sense of innocent delight. We children marveled at the conspiracy of intimacy between God and Grandma.

In the spiritual Body, I picture her as a nerve within the sympathetic nervous system, a sensor that God entrusted with the moment-by-moment responsibility of sensing need. Pastor Smith had sent cries for help to the Head. My grandmother heard the transmitted impulse from the Head and supplied whatever resources were needed.

Grandma had prepared all her life for that behind-the-scenes role. In her youth she had physical energy and beauty. During those busy years of rearing eleven new lives, despite constant demands on her schedule, she had taken the time to know God. She had saturated her mind with the Word of God, storing away in memory whole books of the New Testament, as well as all of the Psalms. Later, when her body grew old and withered, she became a clear channel for God's grace.

In the human body, a minute amount of the proper hormone can guide the transformation necessary to produce new life. In the spiritual Body, the still, small voice of God, when heard and responded to, can change a person, a community, and perhaps a world.

APPARENT DETOURS

In my experience, God normally guides in subtle ways: feeding ideas into my mind, speaking through a nagging sense of dissatisfaction, inspiring me to make a better choice, bringing to the surface hidden dangers of temptation. Only on a few occasions have I felt unmistakable guidance, as if from a direct connection to the Head. The Spirit, a prompter, supplies real help, though in ways that will not overwhelm my freedom.

As I look back on my life, circumstances fit together in a kind of pattern, despite the fact that at various points the opposite seemed true. For example, in my childhood and teenage years I wanted to be a missionary. My parents had impressed on me values that ranked helping the needy as among the highest ends a person could seek. And so, following my father's example, I decided to pursue a career in construction. My father had built schools, hospitals, and homes, and I knew such skills would prove useful in India.

Declining an uncle's generous offer to pay my way through medical school, instead I studied masonry, carpentry, and principles of engineering for four years. After my full apprenticeship I spent one year at Livingstone College taking a medical course on first aid and basic treatments. There, for the first time, I felt a tug toward medicine (mainly because of the experience at Connaught Hospital mentioned earlier), and I fleetingly wondered whether I had made a mistake by turning down my uncle's offer four years before.

Putting these thoughts behind me, I called on the director of my parents' mission and announced to him my willingness to serve in India. To my surprise, he did not see matters in quite the same light. He asked numerous questions about my motives and preparation and then cordially said no. He judged me unready for the kind of work the mission required and suggested more preparation, after which I could submit another application.

I was crushed. God's will had seemed so clear to me, and now this key person was standing in the way.

Following his advice, I signed up for a course at the Missionary Training Colony, a school that taught how to manage life in remote settings. There too I took a brief course in medicine, and the inner voice inside me got louder and louder. I felt inescapably drawn to the field of medicine—so intensely that I withdrew from the two-year missionary course and enrolled in medical school.

At this point, the four years in the building trades nagged at me as a wasteful diversion, a detour. Despite the late start, I did well in school and finished my general training as a physician. Again I presented myself to a mission board, proudly qualified in both construction and medicine. Again I was turned down! This time the interference came from the Central Medical War Committee of Great Britain. They rejected my application to work in a mission hospital and instead ordered me into the bomb casualty services in London. Impatiently biding my time during the forced delay, I studied for qualification in orthopedic surgery.

Twice my good plans had been stymied, once by a wise and godly mission administrator and once by a secular committee of bureaucrats. Each time I felt shaken and confused. Had I somehow misread God's will for my life?

I backed into medicine by apparent accident. Now, as I look back, I can see that God's hand was directing me at every step. Eventually, a man named Bob Cochrane convinced the same Central Medical War Committee to assign me to a new medical college in Vellore, India, and it was he who brought my life's vocation into sharp focus.

REVELATION AT CHINGLEPUT

The author Frederick Buechner gives a succinct summary of discerning God's will: "The place God calls you to is the place where your deep gladness and the world's deep hunger meet."

When the opportunity to return to India finally came, I stipulated a one-year contract, still uncertain about my future. I taught in the Christian Medical College, performed surgery, and filled in daily hospital duties. After a few months I scheduled a visit with my sponsor, Dr. Robert Cochrane.

A renowned skin specialist, Bob Cochrane supervised the leprosy sanitarium in Chingleput, a few miles south of Madras (Chennai). My own hospital did not admit leprosy patients, and I had little familiarity with the disease. Bob showed me around the grounds of his hospital, nodding to the patients who were squatting, stumping along on bandaged feet, or following us with their unseeing, deformed faces. Gradually my nervousness melted into a sort of professional curiosity, and my eyes were drawn to the hands of the patients.

I study hands as some people study faces—often I remember them better than faces. At the leprosarium, hands waved at me and stretched out in greeting. But these were not the exquisite paradigms of engineering I had studied in medical school. They were twisted, gnarled, ulcerated. Some curled into the shape of a claw. Some had missing fingers. Some hands were missing altogether.

Finally, I could restrain myself no longer. "Look here, Bob," I interrupted his long discourse on skin diseases. "I don't know much about skin. Tell me about these hands. How did they get this way? What do you do about them?"

Bob shrugged and said, "Sorry, Paul, I can't tell you. I don't know."

"Don't know!" I responded with obvious shock. "You've been a leprosy specialist all these years and you don't know? Surely something can be done for these hands!"

Bob Cochrane turned on me almost fiercely, "And whose fault is that, if I may ask—mine or yours? I'm a skin man. I can treat that part of leprosy. *You're* the bone man, the orthopedic surgeon!" More calmly, with sadness in his voice, he told me that to his knowledge not one orthopedic surgeon had studied the deformities of the fifteen million people who suffered from leprosy.

As we continued our walk, his words sank into my mind. Leprosy afflicted more people worldwide than the number deformed by polio or disabled by auto accidents—and not one orthopedist to serve them? Cochrane blamed a basic prejudice. Because of the stigma surrounding leprosy, most doctors kept their distance.

A few moments later I noticed a young patient sitting on the ground trying to remove his sandal. His disabled hands refused to cooperate as he attempted to wedge the sandal strap between his thumb and the palm of his hand. He muttered that things were always slipping from his grasp. On sudden impulse I moved toward him. "Please," I asked in Tamil, "may I look at your hands?"

The young man arose and thrust his hands forward. I held them in mine, a bit reluctantly. I traced the deformed fingers with my own and studied them intently. Then I pried his fingers open and placed my hand in his in a handshake grip. "Squeeze my hand," I directed, "as hard as you can."

To my amazement, instead of the weak twitch I had expected to feel, a sharp intense pain raced through my palm. He had a grip like a vise, his fingernails digging into my flesh like talons. I cried out for him to let go and looked up with irritation. Immediately, the gentle smile on his face disarmed me. He did not know he was hurting me—and that was my clue. Somewhere in that severely deformed hand were powerful muscles. They were obviously not working in coordination, and neither could he sense the force he was using. Could those muscles be liberated?

I felt a tingling as if my whole life were revolving around that moment. I knew I had arrived in my place, the place where my deep gladness and the world's deep hunger would somehow meet.

That single incident in 1947 changed everything for me. From that instant I knew my calling as surely as a cell in my body knows its function. Every detail of that scene—the people standing around the sanitarium grounds, the shade of the tree, the face of the patient whose hand I was holding—remains etched into my mind. It was my moment, and I had felt a call of the Spirit of God. I knew when I returned to my base I would have to point my life in a new direction, one that I have never doubted since.

Decades later, I look back with profound gratitude on the time I spent in construction and engineering. Hardly a day goes by that I do not use some of those principles in trying to perfect a rehabilitation device or design a better shoe or apply engineering mechanics to surgery techniques or set up experiments on repetitive stress. And I am equally grateful for the detour that forced me into surgery.

I have stood under the thatched roof of our New Life Center in India and reflected on God's pattern in all those years. As I watch patients do carpentry in our workshop, and the smells of wood shavings and rhythmic sounds of tools rush in, I flash back to my days in a London carpentry shop among my fellow apprentices. I quickly stir from my reverie and see the differences. These are all Indian leprosy patients with reconstructed hands and tools adapted to protect them. God has permitted me the honor of serving them on several levels: as a doctor treating their disease, as a surgeon remaking their hands, and as a carpentry foreman helping to fashion new lives for them.

Only the zigzag course of guidance allowed me to interact with my patients on all these levels. At any point—if I had gone to India earlier, for example, or had bypassed those years in construction—I could easily have strayed slightly out of line and thus have proved less useful. In hindsight, I have a settled sense that God was planning out the details of my life even when the movements at the time seemed like detours. I take great comfort in the promise of Romans 8:28: "Moreover we know that to those who love God, who are called according to his plan, everything that happens fits into a pattern for good" (Phillips).

CHAPTER TWENTY-TWO

GOD'S LIKENESS

The modern ambiance of the Sistine Chapel distorts Michelangelo's original vision for his magnum opus. Visitors to the Vatican enter in groups of several hundred at a time, many of them clasping white plastic headphones to their ears. Instead of looking up when they walk into that splendid room, they follow the trail of red tape on the floor that marks an area where the audio guide is being transmitted.

Nothing can prepare the visitors for what they see when, on cue, they raise their heads. Magnificent works of art cover every inch of the walls and vaulted stone ceiling: the division of light and darkness, the creation of the sun and planets, the days of Noah, the Last Judgment. In the focal center, the calm eye in the swirl of frescoes, Michelangelo has rendered the creation of man.

I linger in that sublime room after most tourists have left. Twilight approaches, and the light has ripened to a golden hue. My neck aches slightly from supporting my head at odd angles, and I wonder how Michelangelo must have felt after a day's work on his scaffold. My eyes drift over to the pivotal scene of God imparting life to the first human, their fingers almost but not quite touching. In a boldly controversial move, the artist did not shrink from portraying God in the image of man. In fact, if you snapped a digital photo of Adam's face, aged it around the eyes, and crowned it with flowing white locks and a beard, you would have Michelangelo's depiction of God the Father.

Can any artist render God? The Old Testament insists that God is spirit and cannot be captured in a graven image; one of the Ten Commandments explicitly forbids it. After living in a country where graven images and idols abound, I understand the prohibition.

Hinduism has thousands of gods, and I can hardly walk a block in an Indian town without seeing an idol or representation. As I observe the effect of those images on the average Indian, I note two results. Most commonly, the images trivialize the gods: they lose any aura of sacredness and mystery and become rather like mascots or good luck charms. A taxi driver mounts a goddess statue on his taxicab and offers flowers and incense as a prayer for safety. At the other extreme, some gods personify powers that evoke a sense of fear and oppression. Calcutta's violent goddess Kali has a fiery tongue and wears a garland of bloody heads around her waist. Hindus may worship a snake, a rat, a phallic symbol, even a goddess of smallpox.

Wisely, the Bible warns against reducing the image of God to the level of physical matter. Any such image limits our understanding of God's real nature: we may begin to think of God as a bearded old man in the sky, like the figure in Michelangelo's painting. The notion of an omnipotent Spirit who spoke the universe into being gets lost. "With whom, then, will you compare God?" asks Isaiah. "To what image will you liken him?" (Isaiah 40:18).

GOD INCOGNITO

Christians believe we got a true and authentic image of God in the person of Jesus Christ. The Spirit took on flesh, a human body of skin and bone and blood and nerve cells. The book of Hebrews describes Jesus as the "radiance of God's glory and the exact representation of his being" (Hebrews 1:3). In other words, if you want to know what God is like, look at Jesus.

Ophthalmologists warn against looking at the brilliance of the sun, even for an instant. Doing so will overwhelm light-receptors and sear the retina like a brand of fire. For thirty-three years Jesus gave us a clear image through which we can perceive God's own self—something like the pinhole camera that allows us to see a solar eclipse without going blind. Here, though, is a strange truth: the image Jesus revealed surprised nearly everyone.

Those of us familiar with the Jesus story may not appreciate the shock, the cataclysmic shock, of God incognito. Jesus missed the people's expectations of God so widely that some asked, incredulous, "Isn't this the carpenter's son?" An ethnic slur followed, "Nazareth! Can anything good come from there?" Even Jesus' brothers did not believe him, questioning his sanity. One of his inner circle betrayed him and another denied him with a curse.

Jesus claimed to be a king greater than David, but little about him suited the image of royalty. He carried no weapons, waved no banners, and the one time he permitted a processional he rode on a donkey, his feet dragging on the ground. To put it bluntly, Jesus did not measure up to the image expected of a king—and certainly not of a God.

We instinctively think of Jesus as a perfect physical specimen, and religious art usually portrays him as tall, with flowing hair and fine features modeled after the accepted ideals of the artist's culture. On what basis? From the evidence, nothing about Jesus marked him as a physical standout.

Once in my childhood my gentle Aunt Eunice came home from a Bible study enraged. Someone had read a description of Jesus, written by the historian Josephus, that characterized him as a hunchback. Aunt Eunice trembled with shame and anger, and her face flushed scarlet. It was blasphemy, she declared. "Utter blasphemy! That is a horrid caricature, not a description of my Lord!" An impressionable child, I could not help nodding in sympathetic indignation.

Although the notion disgusted me at the time, now it would not upset me at all to discover that Jesus was no ideal physical specimen. Although the Bible does not include a physical description of Jesus, there is a description of sorts, in a prophecy of the suffering Servant in Isaiah:

> He had no beauty or majesty to attract us to him,
>> nothing in his appearance that we should desire him.
> He was despised and rejected by mankind,
>> a man of suffering, and familiar with pain.

> Like one from whom people hide their faces
>> he was despised, and we held him in low esteem. (Isaiah 53:2-3)

Furthermore, Jesus identified most closely with those perceived as unattractive and useless. He said of the hungry, the sick, the estranged, the naked, and the imprisoned that whatever we do for one of the least of these people, we do for him (Matthew 25:40). We meet the Son of God not in the corridors of power and abundance, but in the byways of human suffering and need.

In terms of the image the world admires—the image we exploit today in status rankings, beauty contests, and Forbes' lists of the wealthiest—Jesus made no special mark. Yet that one from Nazareth, a carpenter's son, a bruised body writhing on a cross, even he could express the exact likeness of God. I cannot exaggerate the impact of that truth as it fully dawns on a person who will never measure up: a leprosy patient in India, for instance, unspeakably poor and physically deformed. For such a person, Jesus becomes a harbinger of bright hope.

THE SAME MINDSET

I had never grasped the revolutionary pattern Jesus laid down until I began working among leprosy patients. Again and again I saw these people, so cruelly ostracized from society, somehow radiate the love and goodness of God. They had a natural right to bitterness, yet the spiritual maturity among patients who chose to follow Jesus shamed us doctors and missionaries.

As if the image of God that Jesus presented was not shocking enough, the New Testament makes clear that Jesus' followers should express that same image. What he modeled—humility, servanthood, love—become the standard for his Body. Recall the passage from Philippians that says plainly, "In your relationships with one another, have the same mindset as Christ Jesus" (Philippians 2:5).

I search my memory bank for the people I have known who best express this "same mindset." As a child, I often attended large churches and retreat centers where I listened to some of the most famous Christian speakers in England, many of whom demonstrated eloquence and erudition. Instead, another kind of speaker holds a special place in my memory: Willie Long.

I encountered Willie Long in a Primitive Methodist church at a seaside resort. As Willie mounted the pulpit, the fish scales still clinging to his blue fisherman's jersey filled the hall with a pungent aroma. Yet this uneducated man with a thick Norfolk accent, unconventional grammar, and simple faith probably did more to nudge my own faith in those formative years than the entire company of famous speakers. When he stood to speak of Christ, he spoke of a personal friend, and the love of God radiated from him, through his tears. Willie Long, of little consequence in the image of men, showed me the image of God.

Later, in India, I observed with awe the spiritual rapport that bonded patients to the surgeon Mary Verghese. One of my most promising students, Mary suffered a horrific automobile accident that left her paralyzed from the waist down. For months she lay in her hospital bed, resisting physical therapy. She was staking her hopes on divine healing, she said, and rehabilitation exercises for paraplegia would waste her time since one day soon God would restore the full use of her legs.

Ultimately, Mary gained the courage to relinquish that demand for miraculous healing in exchange for the sense of a spiritual power best revealed in her weakness. Against all odds, she completed her surgery requirements and became a dynamic force in the Christian Medical College hospital.

In addition to the paraplegia, Mary had also suffered severe facial injuries. After a series of operations to rebuild the bony infrastructure of Mary's cheeks, the plastic surgeon had no choice but to leave a large, ungainly scar right across her face. As a result, she had an odd, asymmetrical smile. By standards of physical perfection, she did not rate high. Yet she had a singular impact on the patients at Vellore.

Dejected leprosy patients would loiter aimlessly in the hallways of their wards. Suddenly they would hear a small squeak that announced the approach of Mary's wheelchair. At once the row of faces lit up in bright smiles as though someone had just pronounced them all cured.

Mary had the power to renew their faith and hope. Thus, when I think of Mary Verghese, I see not her face but its reflection in the smiling faces of so many others, not her image but the image of God poured through her broken human body.

One last figure towers above all others who have influenced my life: my mother, Granny Brand. I say kindly and in love that my aged mother had little of physical beauty left in her. She had been a classic beauty as a young woman—I have photographs to prove it—but not in old age. The rugged conditions in India, combined with crippling falls and her battles with typhoid, dysentery, and malaria had made her a thin, hunched-over old woman. Years of exposure to wind and sun had toughened her facial skin into leather and furrowed it with wrinkles as deep and extensive as any I have seen on a human face. She knew better than anyone that physical appearance had long since failed her, and for this reason she adamantly refused to keep a mirror in her house.

At the age of seventy-five, while working in the mountains of South India, my mother fell and broke her hip. She lay all night on the floor in pain until a workman found her the next morning. Four men carried her on a string-and-wood cot down the mountain path to the plains and put her in a jeep for an agonizing 150-mile ride over rutted roads. She had made the same trip before, after a head-first fall off a horse, and already had experienced some paralysis below her knees.

Shortly thereafter, I scheduled a visit to my mother's mud-walled home in the mountains in an attempt to persuade her to retire. By then she could walk only with the aid of two bamboo canes taller than herself, planting the canes and lifting her legs high with each painful step to keep her paralyzed feet from dragging on the ground. Yet she continued to travel on horseback and camp in the outlying villages in order to preach the gospel and treat sicknesses and pull the decayed teeth of the villagers.

I came with compelling arguments for her retirement. "Mother, it's unsafe for you to live alone in such a remote place, with good help a day's journey away," I told her. With her faulty sense of balance and paralyzed legs, she presented a constant medical hazard. Already she had endured fractures of vertebrae and ribs, pressure on her spinal nerve roots, a brain concussion, a fractured femur, and a severe infection of her hand. "Even the best of people do sometimes retire when they reach their seventies," I said with a smile. "Why not move to Vellore and live near us?"

Granny Brand threw off my arguments like so much nonsense and shot back a reprimand. Who would continue the work? There was no one else in the entire mountain range to preach, to bind up wounds, and to run the farm and training center. "In any case," she concluded, "what is the use of preserving my old body if it is not going to be used where God needs me?"

And so she stayed. Eighteen years later, at the age of ninety-three, she reluctantly gave up sitting on her pony because she was falling all too frequently. Devoted Indian villagers began bearing her in a hammock from town to town. After two more years of mission work, she finally died at age ninety-five. She was buried, at her request, in a simple, well-used sheet laid in the ground. She abhorred the notion of wasting precious wood on coffins, and she liked the symbolism of returning her physical body to its original humus even as her spirit was set free.

One of my strongest visual memories of my mother comes from a village in the mountains she loved, perhaps the last time I saw her in her own environment. She is sitting on a low stone wall that circles the village, with people pressing in from all sides. They are listening to her talk about Jesus. Heads are nodding in encouragement as she answers their deep, searching questions. Granny's own rheumy eyes are shining, and standing beside her I can see what she must be seeing through failing eyes: intent faces gazing with absolute trust and affection on one they have grown to love.

They are looking at a wrinkled, old face, but somehow her shrunken tissues have become transparent and she is lambent spirit. Granny Brand had no need for a mirror made of glass and polished chromium; she had the reflected faces of thousands of Indian villagers. Her worn-out physical image only enhanced the image of God beaming through her like a beacon.

◆ ◆ ◆

Willie Long, Mary Verghese, Granny Brand—these are three in whom I have seen the image of God most clearly. I do not say that a Miss Universe or a handsome Olympian cannot show forth the love and power of God, but I do believe such a person has a disadvantage. Human self-image thrives on physical attractiveness, social status, intelligence, talent. Any such quality that a person can rely on makes it more difficult for that person to rely on the Spirit of God.

Popularity and acclaim tend to suppress the traits—humility, self-lessness, love—that Christ requires of those who would bear his image. Rather, God's Spirit shines most brightly through the frailty of the weak, the powerlessness of the poor, the deformity of the hunchback. The message is clear: "God has put the body together, *giving greater honor to the parts that lacked it*, so that there should be no division in the body, but that its parts should have equal concern for each other" (1 Corinthians 12:24-25, emphasis added).

As we allow the mindset of Christ to guide us, we too become more accurate bearers of his image. "And we all, who with unveiled faces contemplate the Lord's glory, are being transformed into his image with ever-increasing glory, which comes from the Lord, who is the Spirit" (2 Corinthians 3:18).

CHAPTER TWENTY-THREE

A PRESENCE

To my mind nothing in all of nature rivals the human hand's combination of strength and agility, tolerance and sensitivity. Our finest activities—art, music, writing, healing—depend upon hands. In my surgical career I have specialized in the human hand. Naturally, then, when I think of the incarnation, I visualize the hands of Jesus.

I can hardly conceive of God taking on the form of an infant, yet Jesus entered our world with the tiny, jerky hands of a newborn, with miniature fingernails and wrinkles around the knuckles and soft skin that had never known abrasion or roughness. "The hands that had made the sun and stars," said G. K. Chesterton, "were too small to reach the huge heads of the cattle." Too small, as well, to change his own clothes or put food in his mouth. The Son of God experienced infant helplessness.

Since I have worked as a carpenter, I can easily imagine the hands of the young Jesus as he learned the trade in his father's shop. His skin must have developed calluses and tender spots. He felt pain gratefully, I am sure. Carpentry is a precarious profession for my leprosy patients, who lack the warning of pain that allows them to use tools with sharp edges and rough handles.

Then came the hands of the physician. The Bible tells us strength flowed out of them when Jesus healed people. He chose not to perform miracles *en masse* but rather one by one, touching each person he healed. He touched eyes that had long since dried out, and suddenly they admitted light and color. He touched a woman with a hemorrhage, knowing that by Jewish law she would make him unclean. He touched people with leprosy—people no one else would touch in those days. In small and personal ways, Jesus' hands were setting right what had been disrupted in his beloved creation.

The most important scene in Jesus' life also involved his hands. Those hands that had done so much good were taken, one at a time, and pierced through with a thick spike. My mind balks at visualizing the scene. I have spent my life cutting into hands, delicately, with scalpel blades that slice through one layer of tissue at a time, to expose the maze of nerves and blood vessels and bones and tendons and muscles inside. I have conducted treasure hunts inside splayed hands, searching for healthy tendons to attach to fingers that have been useless for twenty years. I know what crucifixion must do to a human hand.

Executioners of that day drove their spikes through the wrist, directly through the carpal tunnel that houses finger-controlling tendons and the median nerve. It is impossible to force a spike there without crippling the hand into a claw shape. Jesus had no anesthetic. He allowed those hands to be marred and crippled and destroyed.

Later, his weight hung from them, tearing more tissue, releasing more blood. There could be no more helpless image than that of God's Son hanging paralyzed from a tree. "Heal yourself!" the crowd jeered. He had saved others—why not himself? The disciples, who had hoped he was the Messiah, cowered in the darkness or drifted away. Surely they had been mistaken.

In one last paroxysm of vulnerability, Jesus said, "Father, into your hands I commit my spirit." The humiliation of incarnation ended, the sentence served. It was finished.

But the biblical record gives us one more glimpse of Jesus' hands. He makes an appearance in a locked room, where the disciple Thomas is still disputing the story he thinks his friends have concocted. People do not rise from the dead, he scoffs. It must have been a ghost or an illusion. Without warning, unannounced, Jesus enters and holds out those unmistakable hands. The body has changed—it can pass through walls and locked doors. The scars, however, remain, proof that he is the very one they saw crucified.

Jesus invites Thomas to come and trace the scars with his own fingers. Overwhelmed, he says simply, "My Lord and my God!"—the first record of one of Jesus' disciples calling him God directly. Significantly, it's an encounter with Jesus' wounds that sparks the epiphany.

Why did Christ keep his scars? He could have had a perfect body, or no body, when he returned to splendor in heaven. Instead, he kept a remembrance of his visit to earth, and for a keepsake of his time here he chose scars. The pain of humanity became the pain of God.

NEW HANDS

Do we somehow miss the revolution that Jesus set loose? Ancient myths told of the heavens above affecting the earth below. Like kids tossing rocks off highway bridges onto the cars below, the gods dropped judgment and mischief on the earth in the form of rain and earthquakes and thunderbolts. Now the ancient formula has reversed: "As above, so below" becomes "As below, so above." Human actions, such as prayer, affect heaven.

Reflecting on the incarnation, the author of Hebrews notes the progression of intimacy between God and human beings: from the Old Testament style of approaching a distant God through a priest, on to Jesus' up-close visitation. He concludes, "Let us then approach God's throne of grace with confidence, so that we may receive mercy and find grace to help us in our time of need" (Hebrews 4:16). The Head never needs to be awakened or enlightened, and no lack of wisdom or power limits God's activity on earth. The limitation hinges upon member cells obeying the Head in order to serve the rest of the Body.

Today, because Jesus turned over the mission, God's tendrils of activity reach out across the globe. His followers would take the message of grace and compassion and justice to places he never visited during his time on earth, to "Judea and Samaria, and to the ends of the earth." In his life cut short, after all, Jesus had done nothing for most of the world—and that was the plan all along. More than thirty times the New Testament reminds us that we his followers are Christ's Body, the visible presence of God in the world. Where we go, God goes.

I have searched the four Gospels to observe how Jesus prepared for the new phase of headship, and a trend does emerge. During his three years of ministry Jesus gradually turned over his work to his disciples. At first his own hands did the healing, exorcizing, and ministering to needs. As the time of his death neared, Jesus concentrated more on training those who would be left behind. A few key events stand out.

"I am sending you out like lambs among wolves," he warned one of the first groups of his followers to go forth on his behalf (see Luke 10:1-24). Thus he began to entrust sacred tasks to a ragtag group of six dozen novices. Despite the stern warnings, the seventy-two met with great success on their mission, and Jesus responded enthusiastically—I know of no other scene that shows him so full of joy. The work of the kingdom had advanced even as Jesus himself waited, alone.

Later, at the very end of his earthly life, Jesus turned over the entire mission, a transfer that occurred at the Last Supper. "I confer on you a kingdom, just as my Father conferred one on me," Jesus said that night (Luke 22:29). From that point on, he has mainly relied on the self-limiting style of working through human "cells." Clearly, God seems to prefer delegating authority to us humans.

MRS. TWIGG

As a junior doctor on night duty in a London hospital, I called on eighty-one-year-old Mrs. Twigg. Although this courageous woman was battling cancer of the throat, she remained witty and cheerful. In her raspy voice, she asked that we do all we could to prolong her life, and so one of my professors removed her larynx and the malignant tissue around it.

Mrs. Twigg seemed to be making a good recovery until about two o'clock one morning when I got an urgent summons to her ward. She was sitting on the bed, leaning forward, with blood spilling from her mouth. Her face showed an expression of terror. I guessed immediately that an artery back in her throat had eroded. I knew no way to stop the bleeding except to thrust my finger into her mouth and press on the pulsing spot.

Grasping her jaw with one hand, I explored with my index finger deep inside her slippery throat until I found the artery and pressed it shut.

Nurses cleaned up around her face while Mrs. Twigg recovered her breath and fought back a gagging sensation. Fear slowly drained from her as she began to trust me. After ten minutes had passed and she was breathing normally again, with her head tilted back, I tried to remove my finger to replace it with an instrument. But I could not see far enough back in her throat to guide the instrument, and each time I removed my finger the blood spurted afresh and Mrs. Twigg panicked. Her jaw trembled, her eyes bulged, and she gripped my arm fiercely. Finally, I calmed her by saying I would simply wait, with my finger blocking the blood flow, until a surgeon and anesthetist could be called in from their homes.

We settled into position. My right arm crooked behind her head, supporting it. My left hand nearly disappeared inside her contorted mouth, allowing my index finger to apply pressure at the critical point. From visits to the dentist I knew how fatiguing and painful it must be for tiny Mrs. Twigg to stretch her mouth wide enough to surround my hand. Yet I could see in her intense blue eyes a resolve to maintain that position as long as necessary.

With her face a few inches from mine, I could sense her mortal fear. Even her breath smelled of blood. Her eyes pleaded mutely, "Don't move—don't let go!" She knew, as I did, that if we relaxed our awkward posture, she would bleed to death.

We sat like that for nearly two hours. Her imploring eyes never left mine. Twice during the first hour, when muscle cramps painfully seized my hand, I tried to move to see if the bleeding had stopped. It had not, and as Mrs. Twigg felt the rush of warm liquid surge up in her throat, she coughed and grasped my shoulder like a vise.

I will never know how I lasted that second hour. My muscles cried out in agony. My fingertips grew numb. I thought of rock climbers who have held their fallen partners for hours by a single rope. In this case the cramping, four-inch length of my finger, so numb I could not even feel it, was the strand that kept life from falling away. I, a junior doctor in my twenties, and this eighty-one-year-old woman clung to each other super-humanly because we had to—her survival demanded it.

The surgeon came. Assistants prepared the operating room, and the anesthetist readied his chemicals. Orderlies wheeled Mrs. Twigg and me, still entwined in our strange embrace, into surgery. There, with everyone poised with gleaming tools, I slowly eased my finger away from her throat. For the first time I felt no gush of blood. Was it because my finger no longer had sensation? Or had the blood finally clotted after two hours of pressure?

I removed my hand from her mouth and still Mrs. Twigg breathed easily. Her hand continued to clutch my shoulder and her eyes locked on mine. Gradually, almost imperceptibly, the corners of her bruised, stretched lips curled slightly up, forming the hint of a smile. The clot had held. She could not speak—she had no larynx—and she needed no words to express her gratitude. She knew how my muscles had suffered; I knew the depths of her fear. In those two hours in the slumberous hospital wing, we had become almost one person.

WHERE IS GOD?

Forty years later, as I recall that night with Mrs. Twigg, it stands as a kind of parable of the conflicting strains of human helplessness and divine power within each of us. During that agonizing night, my medical training counted very little. What mattered was my presence and my willingness to respond.

Along with most doctors and health workers, I often feel inadequate in the face of real suffering. Pain strikes like a tsunami, with sudden devastation. A woman feels a small lump in her breast, and fear rushes in. A child is stillborn and, for the parents, life itself seems to stop. A young boy is thrown through the windshield of a car; his consciousness flickers on and off like a faulty switch—doctors, ever cautious, offer little hope of recovery.

When suffering strikes, those of us standing close by are flattened by the shock. We fight back the lumps in our throats, make visits to the hospital, mumble a few comforting words, perhaps look up advice on what to say to the grieving. And yet when I later ask patients and their families, "Who helped you most?" I get an unexpected answer. They rarely describe a person with a smooth tongue and a sparkling personality. Instead, they tell me of someone quiet, who listens more than talks, who offers practical help when needed, who does not judge or even offer much advice.

"A sense of presence," they say. "Someone there when I needed her." A hand to hold, a sympathetic, bewildered hug. A shared lump in the throat. Confronted with another's suffering, we long for formulas as precise as the techniques I study in my surgery manuals. But the human psyche is far too complex for a manual. Sometimes the best we can offer is to be there, to love, and to touch.

I have written of lessons from the spiritual Body: the need to serve the Head faithfully, the softness and compliancy of the skin, the diversity of member cells and the marvels that result from their cooperation. Taken together, these provide a sense of presence to the world—God's presence.

When Jesus departed, he transferred that presence to the bumbling community of followers who had largely forsaken him at his death. *We* are what Jesus left behind. He did not leave a book or a doctrinal statement or a system of thought; he left a visible community to embody him and represent God to the world. The seminal metaphor, Body of Christ, could only arise after Jesus Christ had left the earth.

The apostle Paul's great, decisive words about that Body appear in letters addressed to congregations in Corinth and Asia Minor, churches that in the next breath he assails for their faithlessness. Note that Paul, a master of simile and metaphor, does not say the people of God are *like* the Body of Christ. In every passage he says we *are* the Body of Christ. The Spirit has come and dwelled among us, and the world knows an invisible God mainly by our representation, our "enfleshment," of God.

Three biblical symbols—God as a glory cloud, as a Son subject to death, and as a Spirit melding together a new Body—show a progression of intimacy, from fear to shared humanity to shared essence. Where is God in the world? We can no longer point to the holy of holies or to a carpenter in Nazareth. *We* form God's presence through the indwelling of God's Spirit. It is a heavy burden.

I show you a mystery: "In him you too are being built together to become a dwelling in which God lives by his Spirit" (Ephesians 2:22). We bear God's image on this planet.

After World War II German students volunteered to help rebuild a European church that had been destroyed by bombs. As the work progressed, debate broke out on how best to restore a large statue of Jesus with his arms outstretched and bearing the familiar inscription "Come unto Me." Careful patching could repair all damage to the statue except for Christ's hands, which had been destroyed by bomb fragments. Should they attempt the delicate task of reshaping those hands?

The workers reached a decision that still stands today. The statue of Jesus has no hands, and the inscription now reads "Christ has no hands but ours."

Teresa of Avila said it best:

> Christ has no body now but yours. No hands, no feet on earth but yours. Yours are the eyes through which he looks compassion on this world. Yours are the feet with which he walks to do good. Yours are the hands through which he blesses all the world. Yours are the hands, yours are the feet, yours are the eyes, you are his body. Christ has no body now on earth but yours.

DISCUSSION GUIDE

*Joannie Degnan Barth
and Bridget Woltman*

THROUGHOUT THIS BOOK, Dr. Paul Brand and Philip Yancey invite us to "pause and wonder at what God made: the human body." We offer this guide as a challenge to go even further. Within a discussion group, with a reading partner, or even as an individual, you can use these questions to increase mindfulness about the Body of Christ. The set of questions after each section will prompt deeper thought and reflection, and also suggest opportunities for action.

The Hebrew word translated "fearfully" calls to mind reverence and respect. It is our desire that this book and the ensuing discussion will prepare you to better recognize, revere, and reflect on God's presence in the world—in your body and in *the Body*. As Philip Yancey states in the preface, Dr. Brand provided him the "assurance that the Christian life I had heard in theory can actually work out in practice." In your hands, this assurance continues.

PART ONE: IMAGE BEARERS (CHAPTERS 1–2)

1. God creates each of us with a physical body, which Dr. Brand describes as a community made up of individual cells. In the same way that our individual healthy cells work together for the good of the physical body, we each have specific abilities that contribute to the overall Body of Christ.

   Take an honest look at your role in this community. Share examples of how you have served as a loyal member of the Body, supporting the needs of others.

2. Read Psalm 139:13-16:

> For you created my inmost being;
> you knit me together in my mother's womb.
> I praise you because I am fearfully and wonderfully made;
> your works are wonderful,
> I know that full well.
> My frame was not hidden from you
> when I was made in the secret place,
> when I was woven together in the depths of the earth.
> Your eyes saw my unformed body;
> all the days ordained for me were written in your book
> before one of them came to be.

You are an intention of God, who designed and knew you before you entered the world. Talk about your experience of being *known*; describe what this looks like in a few specific settings or relationships, such as your family or church or perhaps a sports team.

How does this compare with times when you feel unnoticed or insignificant?

Thinking of your own experience, how can you make a difference in the lives of others, offering a greater sense of value and significance?

3. Dr. Brand reacted with a start when one of his interns unwittingly conveyed a likeness of Brand's former professor, who the student had never met. We also function as a link in this human chain, passing along expressions, gestures, and other unique manifestations of our character. Do you have a similar story of observing someone's imprint among those in your community?

Consider 2 Corinthians 3:18: "We all, who with unveiled faces contemplate the Lord's glory, are being transformed into his image with ever-increasing glory, which comes from the Lord, who is the Spirit." If you could choose, what characteristic of God would you most want to pass along to family, friends, or close coworkers?

How have you seen evidence of the likeness of God in others?

4. Discuss the Quasimodo effect. How have you encountered the stereotype of equating ugly with bad, and beautiful with good?

Examine your own story—when have you judged and labeled people because of external appearances?

What do you find most difficult about trying to look past the outer shell in order to see the true person inside?

5. Consider this quote from Peter Foster, the RAF pilot whose face was disfigured from burns: "As you prepare to enter the world with your new face, only one thing matters: how your family and intimate friends will respond." Share a story to illustrate how the gaze or opinion of someone else has affected you or defined you.

Does your own physical self-image influence what you believe is true about God's image of you?

Use this self-reflection to become more aware of how others may struggle with this. We are God's reflection in the world, and we have the opportunity to become the mirror for those we encounter. List a few specific acts of affirmation you can employ to demonstrate God's acceptance and love.

6. Dr. Brand maintains that love requires direct contact, often involving physical touch. God did not remain distant but rather came alongside us in the person of Jesus, who then entrusted his followers with the mission of conveying the good news of God's presence to a needy world. Spend some time considering who you can come alongside—at work, in the neighborhood, at church. Address honestly what might be keeping you from these actions.

Reflect: "Now you are the body of Christ, and each one of you is a part of it" (1 Corinthians 12:27). The human body contains around thirty-five trillion cells, an amount five thousand times greater than the number of people on our planet. Respond in prayer to God, reflecting on what you have learned, and how you have been challenged regarding your role in the Body.

PART TWO: ONE AND MANY (CHAPTERS 3–6)

1. The cells in our bodies display an amazing array of specialization, with unique tasks allocated to specific cells. For specialization to work, the

individual cell must forgo all but one or two of its abilities. Let's apply this concept to the Body of Christ. What do you recognize as your own potential area of specialization?

Jesus cautioned his followers to "count the cost" when accepting his invitation to serve others. What is the possible cost of focusing on your specialized service to the Body of Christ (or the cost of a role you are considering)? What reservations might you have, and how could you overcome them?

2. Reflect on the story of José who had lost all contact with the outside world until his sight was surgically restored. He celebrated being reconnected to the community by flashing his smile. Is there a specific connection you hunger for? What would it take to achieve it?

Similarly, John Karmegan endured a lifetime of being judged on his appearance. A simple gesture of acceptance gave John what he had been thirsting for—the sign that he belonged. Share some instances of how you have been included by others, and how you might extend that kindness to someone else.

Although leprosy is rare in developed societies, we have our own groups who feel excluded or rejected based on things such as race, gender, religion, physical or mental disability, age, sexual orientation, economic status, or ethnicity. How can we do a better job of noticing and responding to those who are on the "outside"?

3. Do you ever struggle to believe that you matter? Think back to the 1946 Christmas classic *It's a Wonderful Life* in which an angel is sent from heaven to show a desperate man how his community would have suffered had he never existed. As Dr. Brand reminds us, we matter most in relation to the whole. How do you know that your participation in the Body of Christ makes any difference?

Have you ever seen evidence of your effect on the community?

It may be uncomfortable, but try to share one of your most valuable traits. How might it inspire or empower someone else to improve their relationship with God?

Within your group, take turns noting the faithful qualities you observe and respect in each other. Make a specific plan to affirm someone you know beyond the group, as well.

4. Dr. Brand compares the advantages of autonomy (the amoeba) with the advantages of specialization (the white blood cell). By nature, do you tend to find more pleasure in individual achievement or in working within a group?

In either situation, how do you experience the satisfaction of elevating the whole Body, what Dr. Brand calls the "ecstasy of community"?

5. Read Jesus' prayer in John 17:20-21: "My prayer is not for [believers] alone. I pray also for those who will believe in me through their message, that all of them may be one, Father, just as you are in me and I am in you. May they also be in us so that the world may believe that you have sent me." Dr. Brand contends that unity begins not with our similarity but with our diversity. Where have you seen a situation where differences are affirmed, rather than allowed to obstruct unity in the church?

Think of examples from church history, stories of public figures, accounts retold in this (or any) book, or your own church experience. What positive impact have you seen from embracing diversity?

6. We have learned from Dr. Brand about the upside-down values of God's kingdom: the more we reach out beyond ourselves, the more we are enriched, and the more we then grow in likeness to God. The apostle Paul quotes Jesus as saying, "It is more blessed to give than to receive." Have you found this to be true? If so, describe an experience when you gave of yourself to another and ended up gratified and enriched.

Reflect: "Therefore, I urge you, brothers and sisters, in view of God's mercy, to offer your bodies as a living sacrifice, holy and pleasing to God—this is your true and proper worship" (Romans 12:1). A vast network of blood vessels and neurons connects every cell in the physical body. In light of your connection to the larger Body of Christ, respond to God with a prayer. Meditate on how you can best serve the whole as one of its "cells."

PART THREE: OUTSIDE AND INSIDE (CHAPTERS 7–11)

1. Our skin is described as compliant: able to mold to another shape, protecting what it covers, and embracing what is grasped. Describe a situation in which you observed a particular act of compliance.

   What does it look like when someone puts aside their own comfort to adapt to someone with a distinctly different faith, culture, or personality?

   This type of servanthood takes resolve and practice. Think of someone who is particularly abrasive, and explore ways you can offer a softer touch. How might you better demonstrate the loving embrace of God?

2. Jesus set a high standard with his ministry, and we are encouraged to follow his example. He demonstrated love to one person at a time, face to face, hand to hand. We studied the value of direct contact in Part One, but what about the risk, the potential hazards of personal contact? Have you experienced a negative response when reaching into someone's life? Did it cause you to retreat? Explain what happened.

   Rather than becoming deterred by these risks, discuss how your outlook and preparation can keep such risks from preventing your participation in God's work.

3. In a vivid metaphor of grace, Dr. Brand illustrates how a healthy body responds to the pain of a wounded part. As members of the Body of Christ, we are given this same responsibility. Have you ever been able to support someone else who was serving your church community or local ministry? How did you become aware of his or her needs?

   Discuss a few methods to increase mindfulness about the burdens others carry.

4. Dr. Brand identifies touch as his most precious diagnostic tool, a skill perfected after a great deal of practice. Have you been able to get "in touch with" the needs of someone close to you? Did you (or can you) develop an improved awareness of how to help? How?

   Genuine concern generates greater trust. Identify a relationship in which you would like to offer assistance, but first need to earn trust. List any specific steps you might be able to take.

5. Bones do not burden or restrict us; they support us and enable us to move. God's rules are designed like bones, to provide a strong foundation and framework for our faith. Give an example of how an unchanging principle can provide freedom for a believer.

   Just as stress stimulates bone growth, our challenges often leave us stronger. Share a situation when you felt pushed to the limit and emerged with greater faith?

6. Have you ever endured, or are you currently facing, an episode of doubt? Do you know someone whose faith has been shaken? Dr. Brand reminds us that it is helpful to identify the most basic skeleton of your belief. What can you trust to be true? How can these core beliefs help you or others in times of doubt? Consider the example of the shaky bridge. How can the leadership of those who have passed before you provide the courage to move forward?

   Recount times when you have found God to be faithful, and list these incidences. Rely on this knowledge to assure you in wobbly times.

Reflect: "Indeed, the very hairs of your head are all numbered. Don't be afraid; you are worth more than many sparrows." (Luke 12:7). Our skin cells live only two or three weeks; we lose and regenerate several million of these cells each day. Respond in a prayer of gratitude for the intimate care God promises each member of the Body—including you—and ask for the faith to trust that loving care.

PART FOUR: PROOF OF LIFE (CHAPTERS 12–15)

1. Jesus' short time on earth included many acts that still serve as symbols for us today. Before being crucified, he shared one final meal with his followers. That evening, Jesus applied a new meaning to the traditional Passover meal, instructing us to repeat his actions, to remember him (Luke 22:19). What has been your traditional experience of the Lord's Supper (also known as Eucharist, Holy Communion, Breaking of Bread, and many other terms)?

   Has the analogy offered by Dr. Brand expanded your understanding? What are some new applications you may have gleaned?

Explain how this ceremony can offer meaning, in the past, present, and future.

2. Blood serves as a cleansing agent in the body. Apply this function of blood to what we've learned about forgiveness. Dr. Brand explains that repentance works when each cell "willingly avails itself of the cleansing action of blood." Do you ever struggle to accept or appreciate the liberating gift of forgiveness?

   Just as toxins impede our physical health, even causing us pain, how does sin interfere with our spiritual state?

   Please share an example of how repentance can benefit us, removing contaminants from a lifestyle or relationship. (Use your personal story, if you feel comfortable.)

3. Dr. Brand describes the process of immunizations. How does this help illustrate the power of Jesus' blood to overcome evil in the world?

   Does this knowledge equip you to overcome your own temptations, as it did for Dr. Brand? In what ways? How can this serve as a reminder to rely on the strength of Jesus rather than your own ability, when facing troubles?

   To help this concept sink in, use your own words to restate how Jesus overcame evil.

4. Just as breath sustains the life of our bodies, our Christian faith is sustained by the power of the Spirit. Do you notice a difference in your life when you actively invite the Spirit to lead you? What about when you pause to become aware of God's will, inspired in you? Take time to reflect on how you can improve this practice. Share with others any insights you have found to be helpful.

5. Let's turn to muscles and review how they operate. Like a digital circuit, they respond to a single on-off command to contract. Motion involves an intricate system of levers (bones) and counterbalancing muscles. In addition, muscles must be used (exercised); if an injury or illness prevents motion, the muscles atrophy. Use similarities you recognize in your own spiritual life to draw comparisons within the Body of Christ.

Give examples of growing stronger spiritually with use, or weaker with neglect.

6. When a muscle contracts on its own, it is called spastic—not a malfunction but a rogue action with disregard for the rest of the body. Considering the parallel within the Body of Christ, discuss the slippery slope of doing "good things" for the wrong reason.

How can we be slowly seduced into self-serving behaviors?

Dr. Brand points out the need for special grace to resist egocentric tendencies and notes the power through which God "gently communicates to us." What is your experience with this power?

Reflect: "In this world you will have trouble. But take heart! I have overcome the world" (John 16:33). During a typical lifetime, a human heart beats over 2.5 billion times, pumping over two hundred million liters of nourishing blood. Respond with prayer for strength as we face troubles in this world, and with thanksgiving for the cleansing power of forgiveness when we fail.

PART FIVE: THE LANGUAGE OF PAIN (CHAPTERS 16–18)

1. Pain functions like a language, speaking in a wide range of intensity and volume to alert the body of danger. As Dr. Brand explains, our pain messages trigger important responses throughout our bodies. Thinking now of the pain and peril we encounter in our communities and relationships, describe some signals you have identified as a call for help (including those you use).

Dr. Brand tells us that limping indicates one type of successful adaptation: to avoid to avoid further leg or foot injury. Where have you found the need to adapt—to limp—when a situation becomes painful?

How might you encourage others to do the same?

How well do you recognize and respond to the "limping" you encounter?

2. As we've seen in the illustrations of patients with leprosy, when the body's pain system fails to send warnings, the damage can be

permanent. Recall the story of Sadan, who nearly lost his hands because he lacked that pain signal. Applying this principle to our role in
the Body of Christ, how can we protect from this damage?

Offer any personal stories of when an early-warning system may
have failed or was ignored. What damage resulted? Include thoughts
on how this damage could have been avoided.

3. Although it may seem paradoxical, pain serves a vital role in uniting
the cells of the body. Our cells cooperate to send and react to the alarm
of pain, urgently demanding attention and rerouting resources as
needed. When this pain message fades, certain areas can become detached, and we can disregard the neediest members. How does this
apply to our connection with members of Christ's Body, locally and
across the globe? Give specific examples.

What does this imply regarding our response to those who are
suffering? How could pain and suffering serve to unify the church?

4. Read 1 Corinthians 12:24-26: "God has put the body together, giving
greater honor to the parts that lacked it, so that there should be no
division in the body, but that its parts should have equal concern for
each other. If one part suffers, every part suffers with it; if one part is
honored, every part rejoices with it."

The apostle Paul recognized that a healthy spiritual Body functions
much like a healthy physical body. As members, we are instructed to
share the pain of all other members, to feel the pain of the weakest
part. We are invited to become the tangible presence of God, especially
in times when those suffering feel separated from God's love. Have you
observed a time when the stronger members of the Body became an
expression of the love of God? Explain.

Closer to home, tell a story of a time when you were in a particularly painful situation—how did others attend to you? Was it enough?
Did you feel a sense of God's concern for you?

5. We face risks when tending to the needs around us. For some, re-
peatedly responding to the most urgent crisis will lead to burnout
from overexertion. For others, being inundated with images of

suffering will cause them to become numb or disinterested. It is also possible to lose motivation when a solution to a huge problem does not seem possible. Take heart. God has not called us to resolve every painful situation we see. God has, however, given you a particular interest, some personal expertise, and a unique opportunity. Search your heart and describe an area of need that you are particularly drawn to.

How can you make a difference in an area that might not be attracting much attention? What is the value of smaller contributions over an extended period of time—say, to a needy person in your neighborhood or congregation?

6. Within the Body of Christ, just as in all groups of associated people, conflict and irritation are bound to happen. Dr. Brand suggests that we observe the remarkable way our joints are designed and use this as a source of inspiration for avoiding friction. How would you extend the metaphor?

Think of the roles of various components such as cartilage, canaliculi (channels), synovial fluid; consider the process of lubrication, as well as the amount and distribution of contact. Where does grace become critical, and how would you apply it?

Reflect: "Don't you know that you yourselves are God's temple and that God's Spirit dwells in your midst?" (1 Corinthians 3:16). The friction of a joint such as the knee is only one-fifth that of highly polished metal—about as slippery as ice on ice. In view of what you have learned, pray in gratitude for "the gift of pain" and for the privilege of helping reduce the pain of others.

PART SIX: THE BODY'S CEO (CHAPTERS 19–23)

1. Seven times in the New Testament, Christ is identified as the Head of the church. His leadership style employed the perfect blend of coordination, instruction, and delegation. The functional hierarchy of our nervous system mirrors this structure. Many actions are initiated by reflex, others by direct order, and the "will" of local neurons can

override most muscle activity through "the final common path." Are you aware of any organizations that successfully lead with such a healthy balance? If so, please identify them.

Explore how we can apply the concept of "the final common path" to relationships in our personal lives, our professional positions, and our participation in church (as leaders or members).

2. When on earth, Jesus was the leader of a small group of followers. Now that he has ascended, he serves as Head of a new Body, the church, composed of men and women from all over the world. Talk about this distinction.

As an individual member of the Body of Christ, in what ways can you prepare yourself to listen and respond more attentively to Christ, our Head?

3. God makes his presence known to the world through imperfect people like us, representing all races, sizes, IQs, personalities, and genetic traits. This involves a sort of abdication in which God sets aside omnipotence and adopts a behind-the-scenes role in human history. God's own reputation, as well as that of believers, is tarnished by human failures. The personal conduct of some celebrities professing to be Christian, and even of some church leaders, has contributed to this blemish. Without delving into gossip or specific details, talk about how you deal with reports of this type, and how you think God would want us to react.

What, if anything, can we do to restore the image of God in the world?

How do you respond to this "Divine Trust"?

4. As members of a spiritual Body, we each participate in the marvel of direct contact with the Head. The practice of spiritual discipline is one way to enhance this active relationship with God. Dr. Brand mentions a few ancient and current spiritual disciplines: meditation, fasting, prayer, simple living, worship, celebration, and methodically repeating certain prayers. Are you currently practicing any of these or others? Please share your experience, including what benefit you may have discovered.

Have you adopted any other routines that assist you to hear God's voice and center yourself in God's will? Share something of your personal experience with this intimate relationship.

5. Dr. Brand tells us of the three people in whom he saw the image of God most clearly. What stood out in their descriptions?

Is there anyone in your personal life who reflects the image of God in a way that has profoundly affected you? Take a minute to talk about what you have observed, and how it encourages you.

Make a commitment to share your observations with this person as well. We all need encouragement.

6. We have studied several spiritual lessons regarding the likeness of God in our bodies. The final image is that of the hands of Christ. As his hands, we are entrusted with the mission of reaching out in a variety of ways—to offer comfort, to meet needs, to provide healing—in short, to be the very presence of God. Share any new or refreshed commitment you have made to be an active member of the Body of Christ.

Talk about your willingness to take action, in light of this responsibility we have been given, and address any apprehension you may have.

In your final conversation, take time to sum up the sense of awe and wonder you have gained regarding how we are created, and our trusted position of honor in the Body of Christ.

Reflect: "Do not conform to the pattern of this world, but be transformed by the renewing of your mind. Then you will be able to test and approve what God's will is—his good, pleasing and perfect will." (Romans 12:2). Each of the trillions of cells in your body has direct access to the brain, able to connect and communicate across the brain's hundreds of trillion synapses. Respond with a personal prayer expressing your desire to hear and follow God's will.